CENTRIC 2023

The Sixteenth International Conference on Advances in Human oriented and Personalized Mechanisms, Technologies, and Services

November 13th – 17th, 2023

Valencia, Spain

CENTRIC 2023 Editors

Stephan Böhm, Hochschule RheinMain, Germany

Lorena Parra Boronat, Universitat Politecnica de Valencia, Spain

Printed from e-media with permission by:

Curran Associates, Inc.
57 Morehouse Lane
Red Hook, NY 12571

Some format issues inherent in the e-media version may also appear in this print version.

Printed with permission by Curran Associates, Inc. (2024)

International Academy, Research, and Industry Association (IARIA)
412 Derby Way
Wilmington, DE 19810

Phone: (408) 893-6407
Fax: (408) 527-6351

petre@iaria.org

Additional copies of this publication are available from:

Curran Associates, Inc.
57 Morehouse Lane
Red Hook, NY 12571 USA
Phone: 845-758-0400
Fax: 845-758-2633
Email: curran@proceedings.com
Web: www.proceedings.com

Table of Contents

CENTRIC 2023

Forward

The Sixteenth International Conference on Advances in Human-oriented and Personalized Mechanisms, Technologies, and Services (CENTRIC 2023), held on November 13 - 17, 2023 in Valencia, Spain, addressed topics on human-oriented and personalized mechanisms, technologies, and services, commonly known as I-centric.

There is a cohort of technologies that favored the so called "user-centric" services and applications. While some of them reached some maturity, others are to prove their economics (WiMax, IPTV, RFID, etc). The human-oriented and personalized technologies and services rely on a key set of features, some to be deployed, others getting more mature (personal profiles, preferences, identity, proximity, personal devices, etc.). Following, advanced applications covering human related activities benefit from personalized and human-oriented networks and services, especially preventive and personalized medicine, body networks and devices, or anticipative systems.

The conference provided a forum where researchers were able to present recent research results and new research problems and directions related to them. The conference sought contributions presenting novel result and future research in all aspects of user-centric mechanisms, technologies, and services.

Similar to the previous editions, this event continued to be very competitive in its selection process and very well perceived by the international community. As such, it attracted excellent contributions and active participation from all over the world. We were very pleased to receive a large amount of top quality contributions.

We take here the opportunity to warmly thank all the members of the CENTRIC 2023 technical program committee as well as the numerous reviewers. The creation of such a broad and high quality conference program would not have been possible without their involvement. We also kindly thank all the authors that dedicated much of their time and efforts to contribute to the CENTRIC 2023. We truly believe that thanks to all these efforts, the final conference program consists of top quality contributions.

This event could also not have been a reality without the support of many individuals, organizations and sponsors. We also gratefully thank the members of the CENTRIC 2023 organizing committee for their help in handling the logistics and for their work that is making this professional meeting a success.

We hope the CENTRIC 2023 was a successful international forum for the exchange of ideas and results between academia and industry and to promote further progress in personalization research. We also hope that Valencia provided a pleasant environment during the conference and everyone saved some time for exploring this beautiful city

CENTRIC 20213 Steering Committee

Stephan Böhm, RheinMain University of Applied Sciences - Wiesbaden, Germany
Tzung-Pei Hong, National University of Kaohsiung, Taiwan
Yasushi Kambayashi, Sanyo-Onoda City University, Japan

CENTRIC 2023 Publicity Chair

Lorena Parra Boronat, Universitat Politecnica de Valencia, Spain
Sandra Viciano Tudela, Universitat Politecnica de Valencia, Spain
Jose Miguel Jimenez, Universitat Politecnica de Valencia, Spain

CENTRIC 2023

Committee

CENTRIC 2023 Steering Committee

Stephan Böhm, RheinMain University of Applied Sciences - Wiesbaden, Germany
Tzung-Pei Hong, National University of Kaohsiung, Taiwan
Yasushi Kambayashi, Sanyo-Onoda City University, Japan

CENTRIC 2023 Publicity Chair

Lorena Parra Boronat, Universitat Politecnica de Valencia, Spain
Sandra Viciano Tudela, Universitat Politecnica de Valencia, Spain
Jose Miguel Jimenez, Universitat Politecnica de Valencia, Spain

CENTRIC 2023 Technical Program Committee

Youssef A. Attia, King Abdulaziz University, Saudi Arabia
Duygun Erol Barkana, Yeditepe University, Turkey
Samir Brahim Belhaouari, Hamad Bin Khalifa University, Doha, Qatar
Lasse Berntzen, University of South-Eastern Norway, Norway
Stephan Böhm, RheinMain University of Applied Sciences - Wiesbaden, Germany
Daniel B.-W. Chen, Monash University, Australia
Sabine Coquillart, INRIA, France
António Correia, INESC TEC & University of Trás-os-Montes e Alto Douro, Vila Real, Portugal
Marco Costanzo, Università degli Studi della Campania "Luigi Vanvitelli", Italy
Carlos Cunha, Polytechnic Institute of Viseu, Portugal
Roberto Daza Garcia, Universidad Autónoma de Madrid, Spain
Rui Pedro Duarte, Polytechnic institute of Viseu, Portugal
Luciane Fadel, Federal University of Santa Catarina, Brazil
Rainer Falk, Siemens AG Corporate Technology, Germany
Filipe Fidalgo, Polytechnic Institute of Castelo Branco, Portugal
Alicia García-Holgado, GRIAL Research Group - University of Salamanca, Spain
Stefan Graser, CAEBUS Center for Advanced E-Business Studies | RheinMain University of Applied Sciences, Germany
Till Halbach, Norwegian Computing Center, Norway
Qiang He, Swinburne University of Technology, Australia
Koen Hindriks, Vrije Universiteit Amsterdam, Netherlands
Tzung-Pei Hong, National University of Kaohsiung, Taiwan
Takeshi Ikenaga, Kyushu Institute of Technology, Japan
Imène Jraidi, Advanced Technologies for Learning in Authentic Settings (ATLAS) Lab | McGill University, Montreal, Canada
Christos Kalloniatis, University of the Aegean, Greece
Yasushi Kambayashi, NIT - Nippon Institute of Technology, Japan

Mazaher Kianpour, Norwegian University of Science and Technology (NTNU), Norway
Boris Kovalerchuk, Central Washington University, USA
Alexander Kröner, Technische Hochschule Nürnberg Georg Simon Ohm, Germany
Ravi Kuber, UMBC, USA
Stanislav Mamonov, Feliciano School of Business, USA
Ângela Cristina Marques de Oliveira, Instituto Politécnico de Castelo Branco, Portugal
Pierre-François Marteau, Université Bretagne Sud / Institut de Recherche en Informatique et Systèmes Aléatoires (IRISA), France
Célia Martinie, Université Paul Sabatier Toulouse III, France
José Martins, LIAAD - INESC TEC / Polytechnic of Leiria, Portugal
Thomas Marx, TH Bingen, University of Applied Sciences, Germany
Erik Massarczyk, RheinMain University of Applied Sciences Wiesbaden Rüsselsheim, Germany
David Melhart, Institute of Digital Games | University of Malta, Malta
Pedro Merino, ITIS Software | University of Malaga, Spain
Toshiro Minami, Kyushu Institute of Information Sciences, Japan
Eduardo Miranda, University of Plymouth, UK
Fatma Najar, Concordia University, Montreal, Canada
Areolino Neto, Federal University of Maranhão, Brazil
Khoa Nguyen, Carleton University, Ottawa, Canada
Helder C. R. Oliveira, University of Calgary, Canada
Monica Perusquía-Hernández, NTT Communication Science Laboratories, Japan
Stefan Pickl, Universität der Bundeswehr München, Germany
Melissa Ramos da Silva Oliveira, University of Vila Velha, Brazil
Jagat Jyoti Rath, Instituteof Infrastructure Technology Research and Management (IITRAM), Ahmedabad, India
Valentim Realinho, Instituto Politécnico de Portalegre, Portugal
Ann Reddipogu, RKode Ltd, UK
Michele Risi, University of Salerno, Italy
Armanda Rodrigues, NOVA LINCS | Universidade NOVA de Lisboa, Portugal
José Rouillard, University of Lille, France
Aurora Saibene, University of Milano - Bicocca, Italy
Sandra Sanchez-Gordon, Escuela Politécnica Nacional, Ecuador
Jungpil Shin, The University of Aizu, Japan
Alfredo Soeiro, University of Porto - FEUP, Portugal
Mu-Chun Su, National Central University, Taiwan
Patricia Torrijos Fincias, University of Salamanca, Spain
Carlos Travieso González, University of Las Palmas de Gran Canaria, Spain
Alberto Vergnano, University of Modena and Reggio Emilia, Italy
Christina Volioti, Aristotle University of Thessaloniki / University of Macedonia / International Hellenic University, Greece
Bo Yang, The University of Tokyo, Japan
Hao-Chun Yang, National Tsing Hua University, Taiwan
Christos Zaroliagis, University of Patras, Greece
Alejandro Zunino, ISISTAN-CONICET | Universidad Nacional del Centro (UNICEN), Argentina

Human-Feedback for AI in Industry

Izaskun Fernandez
Intelligent Information Systems
TEKNIKER, member of BRTA
Eibar,Spain
email:izaskun.fernandez@tekniker.es

Kerman Lopez De Calle
Intelligent Information Systems
TEKNIKER, member of BRTA
Eibar,Spain
email:kerman.lopezdecalle@tekniker.es

Eider Garate
Intelligent Information Systems
TEKNIKER, member of BRTA
Eibar,Spain
email:eider.garate@tekniker.es

Regis Benzmuller
IT MES
CONTINENTAL France, Sarreguimes
Sarreguemines, France
email:regis.benzmuller@conti.de

Melodie Kessler
R&D vision industrielle et Deep Learning
CONTINENTAL France, Sarreguimes
Sarreguemines, France
email:melodie.kessler@conti.de

Marc Anderson
LORIA
Université de Lorraine, Inria and CNRS
Vandœuvre-Lès-Nancy, France
email:marc.anderson@inria.fr

***Abstract*—Artificial Intelligence (AI) offers a wide variety of opportunities to the manufacturing industry. However, there are still gaps and challenges to be solved before it can be successfully applied, with data availability and quality being one of the critical factors. The latter highlights the necessity of developing AI systems that can continually learn (from one or more domains) over a lifetime, starting from limited sets of data. This work presents research done on human reinforced learning approaches on small training data sets of open dynamic environments. Beginning this way allows the development of AI models able to learn over time, while taking advantage of a data driven approach along with a knowledge-based approach considering human-feedback as a key enabler.**

***Index Terms*—Human-feedback, Artificial Intelligence, Data Quality, Data Annotation, User-Centric**

I. Introduction

One of the most significant and challenging open problems in Artificial Intelligence (AI) is that of developing systems that can continually learn (from one or more domains) over a lifetime. Although new approaches are appearing to bridge the gap of continuous learning [1], for many years the dominant Machine Learning (ML) paradigms have adopted isolated learning. The latter runs a ML algorithm on a given dataset to produce a model, without any attempt to retain the learned knowledge and use it in the future. The isolated ML approach, has been very successful, but it requires many training examples, and is only suitable for well-defined and narrow tasks in closed environments. This ideal situation is not common in real industrial environments of the small data regime type, i.e., where the amount of available data is scarce. Data may be scarce not only in terms of a limited volume of data, but also due to some environments having highly unbalanced data or slow dynamics which prevent extracting the underlying pattern from the training data. Moreover, to solve many real industrial problems through ML, temporal sequence and/or sensor data should be dealt with, which implies an additional challenge. Accordingly, there is a clear consensus regarding the importance of the quality of the data for the development and deployment of accurate AI based models [2].

Human-feedback could be a valuable source of information for improving ML models and systems by improving data quality and annotations. Collecting and using human-feedback in ML is not a trivial task however. It requires careful design, implementation, and evaluation of different methods and techniques. Indeed, using human-feedback in ML can be expensive, time-consuming, or impractical, especially for large-scale or complex problems. It can also be noisy, inconsistent, or biased due to human errors, preferences, or motivations. It is crucial to take into consideration these challenges and limitations when using human-feedback in ML.

This work presents an overview of different human-feedback mechanisms for human-oriented AI models reinforcement, explored in various real industrial scenarios within the AI-PROFICIENT project. These mechanisms are based on different types and levels of technologies, but all can be classified according to two main groups: implicit and explicit feedback. Implicit feedback is tied to a user action that they would perform (or not perform) regardless of their desire to influence the results given by the AI. Explicit feedback is when the user performs an action specifically designed to enable them to give feedback to the system. Regardless of the feedback type, all the strategies have been designed and implemented following an ethics by design approach, which has contributed to an efficient and good quality data collection, while also promoting user engagement. The most relevant ethical aspects are also presented in this work as good practises/guidelines to be considered when dealing with human-feedback approaches.

The rest of the article is structured as follows. Section II presents the related work. The different human-feedback approaches for AI in industry are detailed in Section III and the main ethical aspects to be considered are detailed in Section IV. Section V describes a successful implementation and deployment of human-feedback for AI in a real use case scenario. And finally, conclusions and lessons learned are summarized in Section VI.

II. Related Work

Data for AI is recognized as an innovation ecosystem in the European AI, data, and robotics framework [4], and data sharing is a critical enabler for competitive AI solutions. Data spaces are a key element of the European Data Strategy, fostering Secure and controlled environments, necessary for eliminating distrust of companies and people when sharing their data. They ensure that data exchanges take place in a safe and secure manner, and in an interoperable manner. The availability of these large interoperable datasets will help to develope more robust and reliable AI based systems, however, incorporating user knowledge into the system is considered as a complementary and indispensable path to improve the AI systems still further, and set the path to cognitive AI systems.

Researchers are defining new types of interactions between humans and AI generically called Human-in-the-loop [5], although definitions of the term vary quite widely as [6] have shown. According to one view, Human-in-the-loop aims to train an accurate prediction model with minimum cost by integrating human knowledge and experience. This enables a significant data requirement reduction, increases reliability and robustness of the AI, and creates explainable AI systems [7], by making humans more effective and more efficient.

There are different approaches implementing Human-in-the loop during different phases of AI system lifecycles: development involving humans in data preparation (including data processing and labelling) [8]; training through interactive ML approaches [9]; data labelling to get explainable AI systems [10]; and reinforcement-oriented approaches [11] using end-users feedback to adjust the AI system to the target user preferences while keeping the model objective optimum.

But, to create an effective Human-in-the-loop system, it is important: to understand how humans interact with machines and to focus on creating natural and easy to use mechanisms that can be wielded through human interaction. It is also important to avoid high cost strategies for human-feedback collection and to research strategies to translate such feedback into exploitable information for AI systems. Such feedback could be provided by humans consciously and explicitly, or inferred from other actions not necessarily linked to feedback.

III. Human-feedback AI

When implementing AI in the industrial context, there are certain challenges that industrial plants face. Two of the most common challenges of AI adoption in industry are data scarcity and the human reluctance to accept AI systems. Perhaps there is no information management system in place to gather vast quantities of data through a multitude of sensors. Or, perhaps, there are enough sensors in place, but the data has never been recorded, labelled and collected. Whatever the reason, beginning to design and develope AI based models with limited annotated data, even partial data, in the sense that whole cases are not represented, or are represented in an unbalanced way, is a real challenge that should be engaged in order to successfully adopt AI models in industry.

In this work, we focus on the use of human-feedback to overcome this data scarcity and data quality challenge - a challenge which usually impacts negatively in the AI model's accuracy - by generating quality data automatically. We present different paths to gather this human-feedback, making a special effort toward reducing as much as possible the human intervention with feedback intention, and putting in place friendly and natural interfaces to facilitate as much as possible the intervention whenever it is necessary. This minimizes the increase of human workload in providing feedback, and fosters the acceptance of the reinforcement AI approaches.

A. *Implicit feedback*

Implicit feedback is considered the best feedback gathering approach in terms of impact upon human workload, since it is collected from an action, optionally done by a human, that has nothing to do with feedback. For example, let's assume there is an AI model running which suggests an optimal value for a certain process parameter in a given moment. Implementing a workflow that registers if a suggestion given by a model has been adopted or not, matching automatically the suggested and the applied values, can strengthen the initial dataset, and thus the AI-model, when retraining by, for example, a reward-punish strategy, although no explicit feedback has been given for that action.

This automatic data strengthening, however, could be negatively affected by erroneous information if the real intention behind the action triggering the feedback is not considered by measuring the degree of compliance of the human [12] with respect to the action. This also happens if the contextual information is not properly managed.

Thus, it is crucial to include in the implicit feedback managing strategy both intention and context management. There are several works in the literature that have proposed methods to differentiate between genuine and unintentional disagreements in implicit feedback [13]. These methods render the information sufficiently accurate, prevent noise creation in the AI model, and successfully use implicit feedback, while minimizing impact on human workload.

In the context of AI-PROFICIENT, the approach is based on a monitoring system to collect the real value and the context information for ensuring the data quality. This feedback management has contributed especially in increasing the data quality, by correcting potential biased or incorrectly annotated data. According to the different typologies of the AI-based models most associated with parametrization and the status of different agents in production plants, in AI-PROFICIENT two different implicit-feedback strategies have been distinguished:

- Predictive AI-based models: the predicted values are compared automatically with the real value registered by the automatism/agent that affects and creates the reinforcement information, indicating if the prediction was correct or not, and if not including the correct value.
- Recommendation (including optimization) AI-based models: the recommended value will be compared

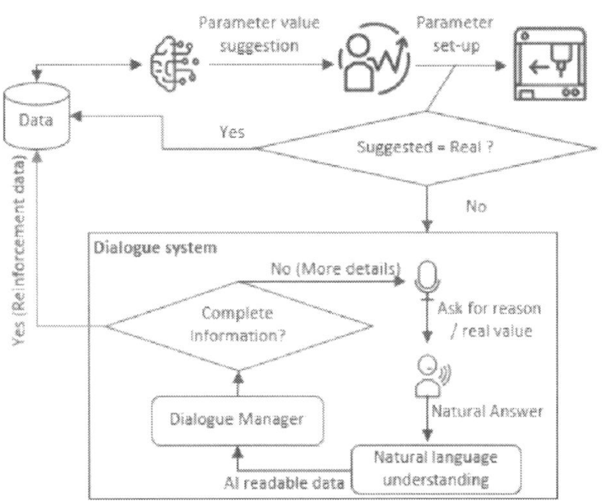

Fig. 1. Natural voice dialogued human-feedback approach.

with the adopted real value introduced in the automatism/agent. When equal it will produce reinforcement positive information and when different, a negative reinforcement information including the recommended and adopted real value.

But in both cases, during the information registration process, contextual information such as product type, timing, and so on, coming from the monitoring system will be registered to ensure data quality assessment.

B. Explicit human-feedback

Reinforcement learning without human intervention is not always feasible. For instance, in a vision recognition scenario, when the AI-based automatic recognition is not the correct one, specific intervention is required to get the right information.

Moreover, explicit human-feedback can overcome implicit feedback drawbacks. In that way, recording the reasons why a decision is taken can make it possible to record contextual information to enrich the model and measure the compliance degree between the human and the AI model.

Since industrial environments are becoming more automated over time, Human-Machine Interfaces (HMI) have increasingly evolved in the last years with the development of new mobile techniques and new gadgets such as smartphones, tablets, or Augmented Reality (AR) glasses. Many solutions have been developed, especially in collaborative robotics, with human-machine interaction capabilities in different degrees, which allow a more intuitive communication with industrial systems: interaction through gestures; programming by demonstration; even dialogue systems, which allow workers to interact with industrial systems in a similar way as they would do with their fellows.

In the following section, different strategies to easily collect explicit human-feedback through advanced Human-Machine Interfaces are presented, taking into consideration the current and future trends in human-machine interaction. This type of feedback could contribute particularly to the improvement of the model by adding to the data collection new previously unobserved types of data which are rarely represented or even unrepresented in the collection, so as to obtain a more balanced data set.

1) Natural voice interaction: The possibility of communicating with industrial systems through natural language is highly encouraged since it triggers acceptance from humans, and dialogue systems are a powerful technological solution to deal with this necessity. Among dialogue systems, task-oriented dialogue systems –as opposed to conversational dialogue systems, which try to emulate regular conversations– are designed to perform specific actions upon a user request and, for this, are especially relevant in industrial contexts. In this sense, task-oriented dialogue systems are powerful technologies that allow workers to work on multiple tasks at once by delegating secondary assignments through communicating with the target system, usually with voice commands. The use of voice instructions to interact with these systems allows workers to use them from a safe distance if necessary, and in a way that they do not need to interrupt their current tasks, leaving the quality of their work unaffected. Furthermore, enhancing these systems with the capability of interacting with users in natural language releases workers from having to learn specific commands to use them.

For feedback gathering purposes, a task-oriented dialogue system should focus on generating the proper dialogues when the AI system fails, asking the human (operator) the necessary questions to obtain the right information and translate it in order to create new insights to reinforce the current model. Fig. 1. shows a generic workflow of this approach.

2) Augmented Reality based interfaces: Augmented Reality (AR) is a technology field that involves the seamless overlay of computer-generated virtual images on the real world, in such a way that the virtual content is aligned with real world objects and can be viewed and interacted with in real time. AR research and development has made rapid progress in the last few decades, moving from research laboratories to widespread availability on consumer devices. Augmented Reality through wearable devices such as Google's Augmented Reality glasses can bring numerous advantages for information visualization, such as displaying relevant information to the driver of a fork-lift through the glasses. Such devices can also be a very useful tool for machinery repair and maintenance, mitigating human errors and possible accidents [14]. Common approaches for AR interaction include tangible User Interfaces (UIs) and free-hand gesture-based interaction. These could be used, not only for advanced information visualization in industrial scenarios, but, in addition, to facilitate the user in providing corrections or suggestions, thus enabling a new channel for the collection of human-feedback for reinforcement purposes.

3) Shop-floor interfaces: New HMIs need to be more sophisticated for enhanced efficiency and remote service operations, especially when workers are interacting with intelligent agents in dusty, humid, or dark, industrial environments. Since operators become involved in the manufacturing process for critical decision-making, the HMI system should allow commands to be easily and rapidly entered in order to increase the accuracy, safety and speed of problem-solving. But not all the industrial scenarios are ready to go for a voice or augmented reality driven HMI.

So, not only the most advanced mechanisms should be considered for feedback management, but extending currently available or newly developed dashboards and interfaces must be also considered to ensure human-feedback gathering in scenarios with different levels of digitalization in terms of interfaces: excel sheets, web-based dashboards, etc. The feedback methods can also benefit from the State-of-the-art techniques applied for uses in the industrial environments, such as active noise suppression, AI-supported and trained Optical Character Recognition, etc.

IV. Lessons Learned from Human-Feedback Mechanisms in Use

In the context of AI-PROFICIENT project, a total of 6 industrial use cases (from Continental and INEOS industrial partners) involving AI models and human-feedback oriented reinforcement strategies have been developed to face different industrial problems. Table I shows a summary of the different human-feedback approaches, according to the classification presented above.

The main common challenge set by the industrial partners in all the scenarios has been not to change the working procedures and to minimize new interfaces and devices in the workplaces for AI models and feedback mechanisms. Taking into account this restriction, the 6 use cases have set up at least one AI model supporting the target problem, and all except one, which has only implemented an explicit human-feedback approach, have combined both, implicit and explicit feedback. Two of them have even combined two different explicit mechanisms.

For implicit feedback, monitoring systems have been set up in the scenarios without affecting existing procedures. While for the explicit feedback, the most common approach has been to extend the interfaces developed for interacting with AI models. For instance, an AR application in INEOS plant that uses a computer vision model to recognize labels has been extended to gather feedback. More specifically, new AR screens have been added to the application enabling users to introduce the correct information by augmented keyboard, and even voice, when the AI model fails. But not only industrial restrictions have been considered. During the design and implementation of the human-feedback mechanisms, an expert team has assessed ethical aspects relative to human acceptance. Following, we list the most remarkable and most often met with aspects that should be considered in order to develope a successful human-feedback AI approach ethically.

TABLE I
OVERVIEW OF NUMBER OF USE CASES IMPLEMENTING FEEDBACK MECHANISMS, PER TYPE

Feedback type	# use cases
Implicit Human-feedback	5
Explicit Human-feedback - Voice	4
Explicit Human-feedback - AR	1
Explicit Human-feedback - Shopfloor HMI	3

- When generating human-feedback based data, regardless of whether it belongs to explicit or implicit interaction, indicate that the data is generated by human-feedback
- Given that voice data, because it is biometric, is inherently sensitive data, when feedback mechanisms make use of it, it is recommended that ethical best practices similar to the measures of Article 5(b) GDPR be implemented, regardless of legal compliance requirements, and that demonstrable consent be obtained.
- In those cases where natural voice and language interaction is part of the human-feedback, take into consideration the operator's mother tongue and if the former is not considered when implementing the mechanism, then, develop a clear, detailed, and practical plan for how the language gap and difficulties related to operator language and HMI use will be bridged.
- The feedback mechanisms should not increase the user workload and if it does, strategies should be implemented in order to try to minimize it for ensuring adoption and acceptance.

V. Human-feedback AI: In Continental Use Case

Following the ethical recommendations and taking into consideration Continental's need to improve a trade blade change process, an AI-based reinforcement approach has been implemented in the Continental plant.

In the beginning of the research, there was no AI-based model deployed in the Continental plant, but at the end of the

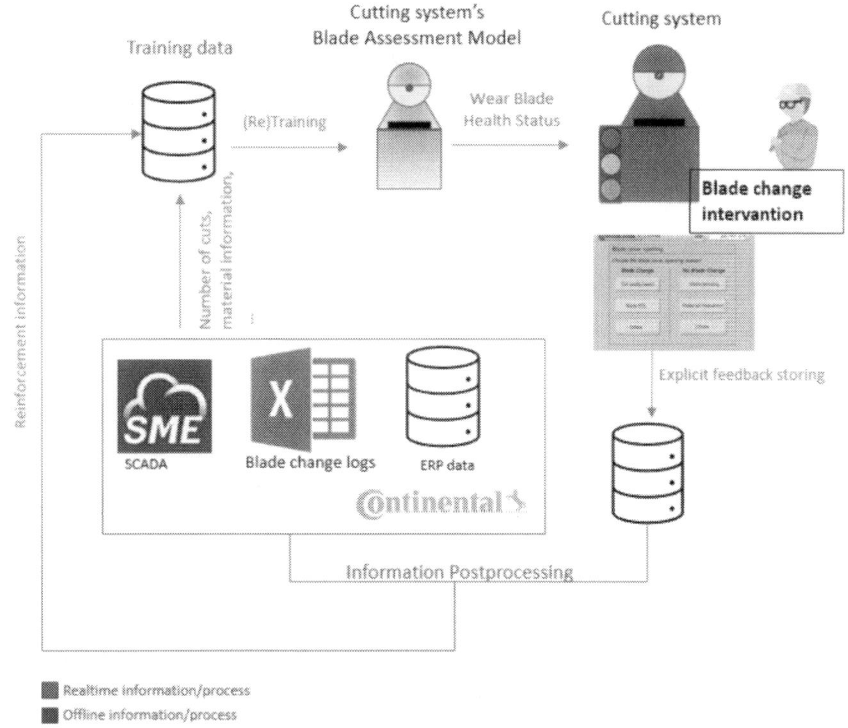

Fig. 2. Continental trade-blade change human-feedback management flow.

workday, the craftsmen and operators used to fill an excel-sheet indicating, among other activities, the blade changes done during the workday, and their approximate time. This information was stored together with the other information associated to the machine with the trade blade, such as material types and compositions, performed cuts, or monitored signals.

Taking advantage of this historical data (from the three last years), necessary actions have been undertaken to develop and deploy an initial AI-based model, aimed at predicting the optimal moment to change the blade, considering all the above mentioned information and the appearance of the blade.

However, due to the imperfection of the AI-model, mainly caused by the scarce and not exact training data, the prediction does not always provide the best moment for the blade change, and so the operator does not always follow the suggestion provided by the model. In order to correct these deviations and train a more robust model as it is used, a human-feedback management approach, based on both implicit and explicit feedback has been implemented and deployed in real plant as summarized in Fig. 2.

A. *Implicit feedback in use*

AI-based model estimation is shown to the operator in terms of a traffic-light visualization, with green indicating a healthy blade status and red indicating the end of life of the blade, and thus a trade-blade change action needed. Currently, adopting the AI-based estimation or disregarding it, when necessary the operator directly performs the blade change or asks the craftsmen to do it.

The current solution in the Continental plant includes in the trade-blade safety-cap, a sensor which records the exact time in the database whenever the cap is opened. The cap can be opened due to different reasons:

- Blade change
- Adjusting of the blade without change
- ...

This in-situ time registry, facilitates an exact time recording but since the cap opening is not necessarily related to a blade change, an explicit feedback mechanism is needed to complement the information and be able to select and use only the correct data for model reinforcement.

B. *Explicit human-feedback in use*

In comparison to the initial situation, the current solution in place in the Continental plant eliminates the operator's need to report blade changes and their approximate time at the end of workday, but on the other hand it requires their intervention each time the monitored blade security-cap is opened to clarify the action reason. The intervention is simplified to a popup in the associated machine HMI, which presents the potential options (see Fig. 3.) that have motivated the cap opening. This pop-up is presented to the operator/craftsmen each time the cap is opened and he/she should only select one of the options. When, and only when, the selected option is related to a blade change action, a data compilation phase is activated,

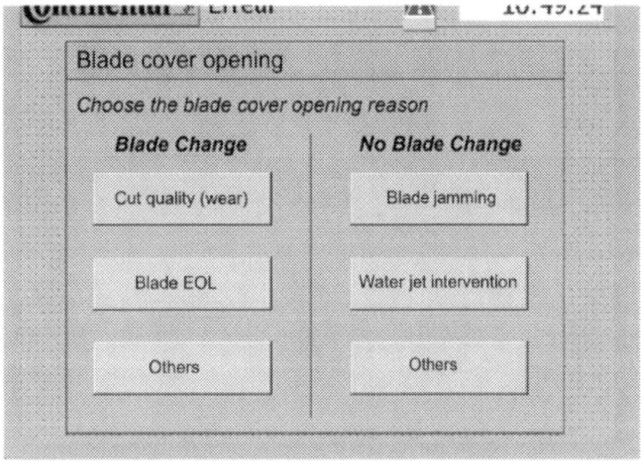

Fig. 3. Continental Shopfloor Interface for trade-blade change explicit-feedback gathering. The language of the operators has been taken into account, so the interface in place is in French, but for better understanding it is translated to English here.

enabling, in real time, the running model to be fed with current real status and so improve further estimations and to update the data to reinforce the model with more accurate data, including a flag indicating the data derived from explicit-human intervention.

The shopfloor HMI was deployed in the Continental plant in 2023-03-15 and is currently working. During this period, the operators have been using it to report the interventions. They have found the mechanism user friendly and no significant increased workload has been identified until now. Fig. 4 shows the annotations that the operators have performed by the explicit feedback mechanism, every time a blade intervention has occurred.

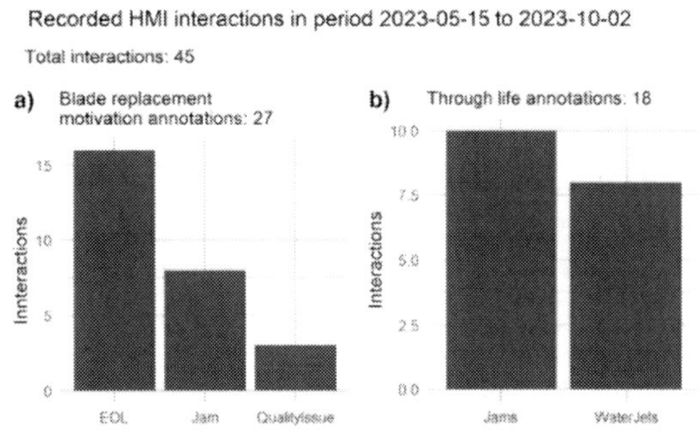

Fig. 4. Summary of annotations provided by users: a) At the removal of the blade b) During the life of the blade.

The reasons that are not related to blade change are also stored and linked to the time they happen but the processing of this data is out of the scope of this work, and will be part of further work. In the period of the 5 months, to date, that the HMI has been in place, a total of 18 interactions related to blade maintenance without blade change actions, have been registered as depicted in the graphic b in Fig. 4..

The annotated data related to blade changes have been used to retrain the model estimating the blade health status. A complete evaluation is a work in progress, but some initial assessments have already been done. Observing the $Rsquare$ metric used for the validation of the fitting[1] of the AI model, the inclusion of the 27 new datapoints obtained through the operator-feedback reporting blade change reason (see details in Fig. 4. a) has improved the value from 0.9683 to 0.9880. This improvement, although small, could be significant on the long run and confirms the human-feedback value for reinforcing AI models.

VI. Conclusions and Further Work

This work presents the research done in exploring the implementation in industrial environments of a reinforcement AI approach by human-feedback. Different technological approaches have been adopted in different industrial scenarios involving AI-based models, solving different real problems, reinforced by implicit and explicit human-feedback gathering workflows. Although the solutions make use of very different technologies, starting from the adaptation of traditional shop-floor interfaces to more sophisticated augmented reality or voice based approaches, human empowerment should be at the center of all of them to ensure a successful adoption of the solution. Accordingly, using the native language of the target industrial scenario, and taking care not to increase the workload of the humans involved, are some of the critical aspects that should be taken into account during the design phase of the Human-feedback AI approach to ensure acceptance and adoption from both the industrial and human side. Furthermore, for industrial acceptance, a non-intrusive improvement of interfaces and mechanisms in place is preferred. Such an approach enables the capture of those (explicit/implicit) operator interactions which directly affect the accuracy of the model, in order to improve data quality, without modifying - or only slightly modifying - the current procedures in the plants.

Further work includes a quantitative evaluation of the impact of the reinforcement approach, with regard to which initial evaluations show promising results, as well as exploring new paths in the human-feedback workflows inline with the identified human and industrial restrictions.

Acknowledgment

This project has received funding from the European Union's Horizon 2020 research and innovation programme project AI-PROFICIENT under grant agreement no. 957391.

References

[1] L. Wang, X. Zhang, H. Su, and J. Zhu, "A comprehensive survey of continual learning: Theory, method and application," arXiv preprint 2302.00487, 2023.

[2] K. López de Calle, E. Garate, and A. Arnaiz, "Towards a Circular Rotating Blade Wear Assessment Digital Twin for Manufacturing Lines," IFAC-PapersOnLine, vol. 55, n. 2, p. 566, 2022.

[3] Meritxell. Gómez, B. Sierra, y S. Ferreiro, "On the Evaluation, Management and Improvement of Data Quality in Streaming Time Series," IEEE Access, vol. 10, pp. 81458-81475, 2022,

[1]Model fitting is a measure of how well a machine learning model generalizes to similar data to that on which it was trained.

[4] S. Zillner, D. Bisset, M. Milano, E. Curry, C. Södergård, T. Tuikka, et al., "Strategic research, innovation and deployment agenda: AI, data and robotics partnership," 3 ed., 136p., 2020.
[5] X. Wu, L. Xiao, Y. Sun, J. Zhang, T. Ma, and L. He, "A survey of human-in-the-loop for machine learning," Future Generation Computer Systems, vol. 135, p. 364-381, 2022.
[6] M. Anderson and K. Fort, "Human Where? A New Scale Defining Human Involvement in Technology Communities from an Ethical Standpoint," International Review of Information Ethics, 2022.
[7] C. Deng, X. Ji, C. Rainey, J. Zhang, and W. Lu, "Integrating machine learning with human knowledge," Iscience vol. 23, no. 11, Elsevier, 2020.
[8] R. Munro and R. Monarch, "Human-in-the-Loop Machine Learning: Active learning and annotation for human-centered AI," Simon and Schuster, 2021.
[9] N. A. Wondimu, C. Buche, and U. Visser, "Interactive machine learning: A state of the art review," arXiv preprint arXiv:2207.06196, 2022.
[10] M. Johnson, A. Albizri, A. Harfouche, and S. Fosso-Wamba, "Integrating human knowledge into artificial intelligence for complex and ill-structured problems: Informed artificial intelligence," International Journal of Information Management vol. 64, p. 102479, 2022.
[11] S. S. Shuvo and Y. Yilmaz, "Home energy recommendation system (hers): A deep reinforcement learning method based on residents' feedback and activity," IEEE Transactions on Smart Grid, vol. 13, no. 4, p. 2812–2821, 2022.
[12] W. Wang, F. Feng, X. He, L. Nie, and T. Chua, "Denoising Implicit Feedback for Recommendation," Proceedings of the 14th International Conference on Web Search and Data Mining,p. 373-381, 2021.
[13] J. Lie, Y. Ren, and K. Deng, "FairGAN: GANs-based Fairness-aware Learning for Recommendations with Implicit Feedback," Proceedings of the ACM Web Conference 2022, p. 297-307, 2022.
[14] A. Martinetti, M. Rajabalinejad, and L. Van Dongen, "Shaping the future maintenance operations: reflections on the adoptions of augmented reality through problems and opportunities," Procedia CIRP vol. 59, p. 14–17, 2017.

LLM Assisted No-code HMI Development for Safety-Critical Systems

Insights of a Short Impirical Study

Matthias Harter
Faculty of Engineering
Hochschule RheinMain - University of Applied Sciences
Rüsselsheim, Germany
e-mail: matthias.harter@hs-rm.de

***Abstract*—This paper represents the outcome of an empirical study conducted with Large Language Models (LLM) on the question whether or not we can expect current and future Artificial Intelligence to assist engineers in the development of embedded software systems for safety-critical applications. Experiments with GPT-4 and other LLMs suggest that current models are capable of *assisting* developers in part in the design of Human-Machine-Interfaces (HMI) for instruments and displays used in aircrafts without the need to manually write a single line of source code (no-code development) while maintaining the highest level of safety and reliability demanded by authorities and customers. The study does not present generally accepted quantitative measures in answering the question how well suited current language models are for this task, but rather provides a qualitative assessment of the capabilities of state of the art AI. It also sheds a light on the deficiencies of today's LLMs in fully understanding technical systems in depth. Instead of completely replacing human engineers we should rather strongly rely on human-in-the-loop policies for the most critical phase of the progressively automated development process, even with more sophisticated and powerful LLMs on the horizon. It should be noted that providing objective evidence to support the argumentation and the findings in this short paper will be the subject of future work.**

Keywords*—*AI; requirements engineering; safety critical embedded software; model-based software design; automatic code generation.

I. Introduction

This section briefly presents the development process of embedded software systems typically employed these days and problem with AI when used in safety critical applications. It is then argued that AI can still be used sensibly instead of doing without it completely.

A. State of the Art

Outsourcing certain steps in the development process of software systems to the machine assistant has recently gained popularity, since it has shown to be beneficial with respect to coding tasks (e.g., GitHub Copilot). For many years prior to the advent of large language models like Chat-GPT and others, extensive automation of the development process of complex software systems has been the goal of many tools and procedures. For instance, the translation process from the source-code in a high-level programming language to the binary executable program (machine code or object code) with compilers like the GNU Compiler Collection (GCC) is a fairly complex and demanding task and leads to solutions that one could attribute in a certain way to an intelligent agent, even though no AI algorithm or machine learning strategy has been traditionally employed of course. This observation is typically made by students of computers science when asked to thoroughly analyze the results of the compilation process for certain single statements in the C programming language by comparing them to the equivalent in the assembly language. The machine based translation of such statements is sometimes more efficient and elegant than the human counterpart, at least for assembly language beginners. Other tasks in the development process have been automated since many years as well, e.g., documentation generation by extracting information from comments within source code snippets (e.g., Doxygen) and static code analysis to name just a few.

Now that LLMs like GPT-4 [1], CodeLlaMa [2], StarCoder [3] or CodeGen [4] are capable of generating source code for dozens of different programming languages (depending on the model and the training), it seems logical to let the AI do the coding job, at least for small portions of a program. An supposedly increasing number of software engineers is using these models as a starting point for their software projects by diving the task into smaller modules which then get completed by hand in a repeated manner (module by module). Putting everything together and making the whole software system work as expected is certainly something that must still be done by human hand. A future seems possible, in which more and more of this work can be outsourced to AI agents (chat bots specialized in coding), not only for small modules or just as an assistant for code completion.

B. LLMs and the problem with safety

The phenomenon of hallucination (better referred to as "confabulation") raises concerns about reliability and trust into this development process, especially when targeted at safety-critical systems in aircrafts or other machinery that must be 100% safe (e.g., medical devices). For this reason, letting today's and even future AI write the code for safe-critical systems is not advisable, no matter how good they are or how promising the outcome will be. Even using LLM generated code for certain smaller parts of a safety-critical system is not recommended, mainly for practical reasons: The source code must be qualified/certified by authorities like the Federal

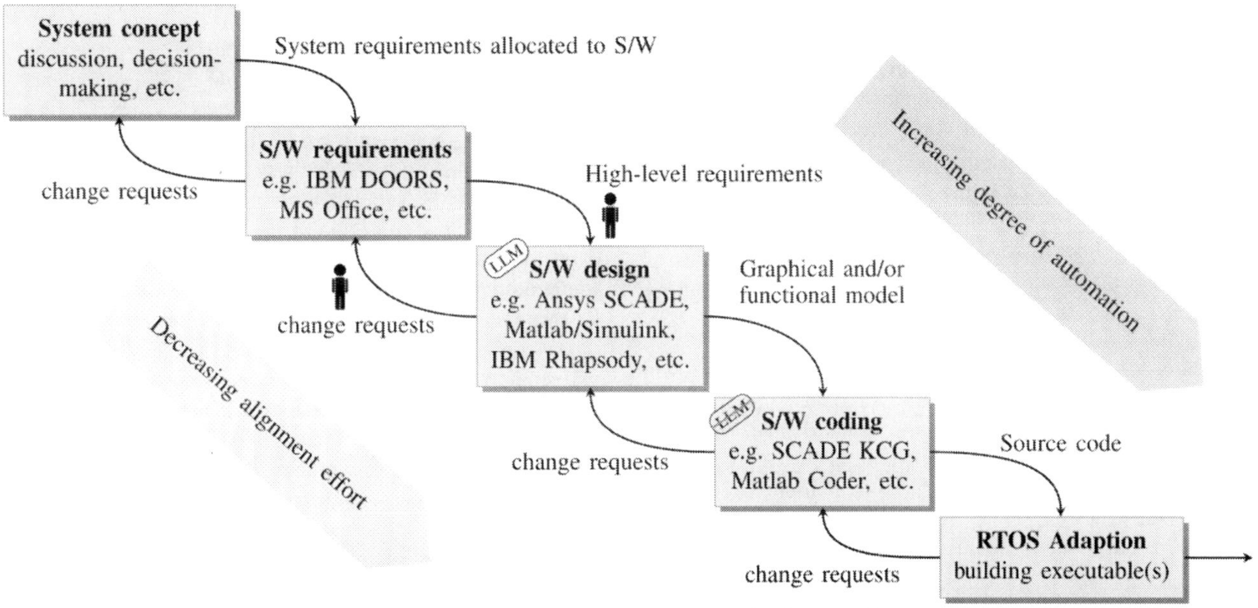

Figure 1. Development process of a safety-critical embedded software system according to DO-178C and ARP 4754A. The process resembles the left side of the famous V-model down to the implementation phase. To the right the process continues with integration, test and verification and validation (not shown).

Aviation Administration (FAA) in the US, the European Union Aviation Safety Agency (EASA) or other specialized institutions acting as certification instances. This certification process is tedious, expensive and ultimately superfluous. Instead of certifying the code every time for every software product or even every version of one and the same software product, we should move from the artifact (the software) to the process and the tools involved. If the tools in the development process are certified to be safe and to comply with safety standards and regulations (e.g., ISO 26262, DO-178C, etc.), we can rely on automation to a greater extent.

Unfortunately, such certification will be hard to achieve in case of LLMs writing source code due to their stochastic nature and the huge amount of training data that defines the capabilities and limits of the tool. The problem of confabulation stems from this design principle. In others word, neither the ever changing code generated by LLMs nor the LLM itself as a tool are good candidates for certification and the target to put trust into. Instead, we can solely rely on the code generation powers of specialized tools, such as Ansys SCADE Suite KCG and SCADE Display KCG, which produce C (ISO-C and MISRA-C [5][6] compliant) or ADA source code. According to the company's website, Ansys SCADE Suite and SCADE Display KCG have been qualified to comply with all relevant safety standards.

C. LLMs at the level of requirements

As argued in the previous subsection, we should not employ LLMs at the coding level in the development process. As their name "large language model" implies, this kind of AI is suited for interpreting natural language and for human-machine conversations (chats). In the development process of embedded software systems natural language plays an important role at the very start of the process. Developers and management discuss the concept of the system, i.e., its capabilities, fundamental properties and limitations. At the end of this step, decisions have to be made and translated into a set of requirements that break down all characteristics of the system. In case of a software system, the requirements are a subset of the complete requirements, of course. In Figure 1, the whole process is depicted with the software requirements being the starting point for a possible integration and adaption of LLMs. The terms "requirements management" and "requirements engineering" represent the underlying tasks and procedures, which can be supported by specialized tools like IBM DOORS or just by simple office software (e.g., Microsoft Office). Most engineering activities in a company on these upper two levels are centered around requirements as more or less formal agreement upon fundamental properties of a system or product, which can get further refined (indicated by the arrows labeled "change request"), if not detailed enough, misleading or ambiguous. Engineers all over the world employ this methodology of requirements engineering as a means to establish a common understanding about the product to be developed. Of course, this understanding is achieved by natural language communication between humans. This step typically takes a significant amount of time and effort and eventually leads to a database of ideally precise, consistent, comprehensive and in terms of technicality detailed phrases in a natural language like English.

At the level of requirements, AI can come into play and support engineers taking the next step. Traditionally it was the developer's job to translate all the requirements into a model of the system, either rather informal as a sketch inside the engineer's brain or workbook or more formal as an architectural schematic, flowchart or state diagram. In the context of safety-critical applications in the mobility sector and other fields, it has become state of the art to use SysML/UML to model the system, using tools like Matlab Simulink, IBM Rational Rhapsody or Ansys SCADE Suite [7][8]. This modeling task can be quite sophisticated or even error-prone, if done by unexperienced engineers, but the hypothesis is that it can be aided by LLMs which play the role of an assistant. In this scenario the LLM is given the requirements as input (together with a certain prompt) and the output is a *first and basic version* of the model of the system (sub-system or module) to be developed. Here, the engineer is still needed not only to supervise this translation process, but also to edit, complete and essentially examine the model itself to see if it is properly aligned to the requirements defined by human. Change requests will still be needed frequently in order to keep humans in the loop and are essential to guarantee a system of check and balances.

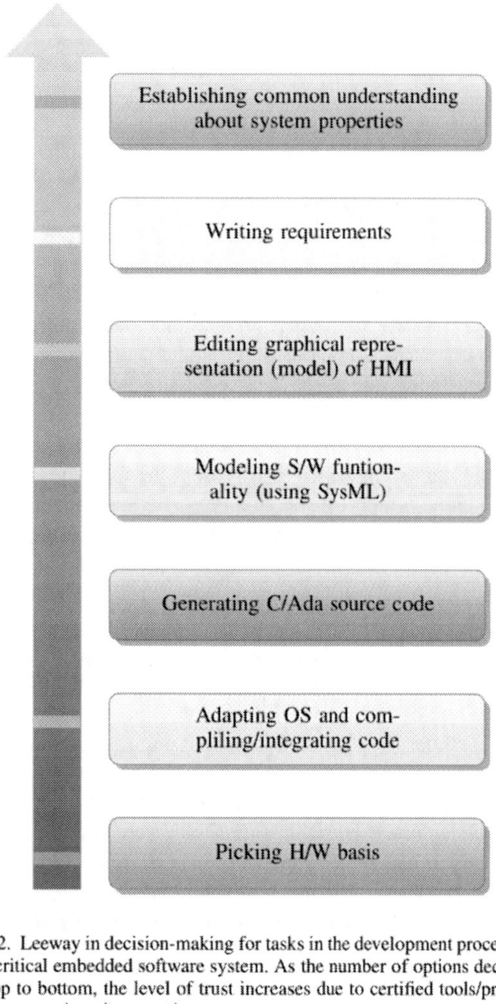

Figure 2. Leeway in decision-making for tasks in the development process of a safety-critical embedded software system. As the number of options decreases from top to bottom, the level of trust increases due to certified tools/products and proven engineering practices.

D. Alignment of automated processes with the human-centric perspective

One can observe a decreasing degree of alignment effort with human-formulated standards and specifications as automation is increasingly incorporated. Generally speaking, mankind should emphasize the importance of (wo)man-machine alignment at the highest level of each development process and will more and more get along with highly automated procedures and tools at the lower end of the process, even for safety-critical sectors. This increasing degree of automation does not mean that we should use LLMs to generate software code, but support us in the design or modeling phase while still keeping control of the model itself. This way engineers can focus on the human-centric viewpoint, i.e., defining what behavior is desirable, and let proven tools like traditional code generators (e.g., SCADE KCG) do the tedious and costly job. This observation is summarized in Figure 2 and further detailed for the use case which was studied for this paper.

At the top level of the whole process humans can choose between a vast variety of options and have to discuss and argue within a group and with the outside world about fundamental characteristics of the product. Writing down the requirements narrows the leeway in decision-making quite a bit, since many options and features turn out to be unfeasible, costly or otherwise undesired. In the case of a system with a Human-Machine-Interface (HMI) or User-Interface (UI) respectively, the requirements get broken down into a model of the graphical representation, which typically consists of a set of basic shapes (primitives like circles, rectangles, text and others) and their properties. This means that, again, many options are omitted and things get further concretized. Modeling the functionality in the next step using SysML is another way of narrowing the leeway. In case of the MBSE approach (Model-Based System Engineering) employed by the SCADE tool family, the SysML model needs to be technically precise and comply with certain modeling principles in order to use the SCADE Code Generator (KCG) for automated code generation. We have the choice between C or Ada and can steer the code generation process to a small extend, but the range of options is rather limited. At the bottom of this diagram we can see that compiling the source code and integrating it into a software ecosystem (usually a real-time OS/RTOS) is a matter of choosing between very few software products (e.g., VxWorks, INTEGRITY-178B or PikeOS).

The same applies to the hardware basis of the system. Obviously, the leeway of decision-making is so narrow that we cannot choose freely between all sorts of hardware and computing platforms, e.g., Raspberry Pi or Arduino. We have to get along with what has been proven to fulfill the highest standards of safety. Specialized hardware offers features and certain safety measures to ensure this (e.g., lockstep mode of operation, majority vote principle, watchdog timers, etc.).

II. Methodology

In this section, the overall scenario is presented and the procedure for investigating the possible applications of AI in the development process of saftey-critical embedded software is explained.

A. Use case

In the scope of this work, the following scenario has been studied: Large language models which have been trained extensively with programming language code (OpenAI GPT-3.5, GPT-4 [1], Salesforce CodeGen [4], StarCoder [3] and CodeLlaMA [9]) are instructed to take the requirements for the display (HMI) of an aircraft instrument as input prompt (together with the system prompt, if applicable) and translate them into a model that is fed into the Ansys SCADE development environment. The model should reflect the requirements as precisely as possible and thus demonstrate the capabilities in understanding natural language from the standpoint of a technical assistant. The aircraft instrument to model is a so-called Primary Flight Display (PFD), which is used by pilots as an indication of the aircrafts attitude in relation to the horizon. This instrument is therefore also called "artificial horizon". It usually also provides information about the aircraft's speed and its altitude (above mean sea level).

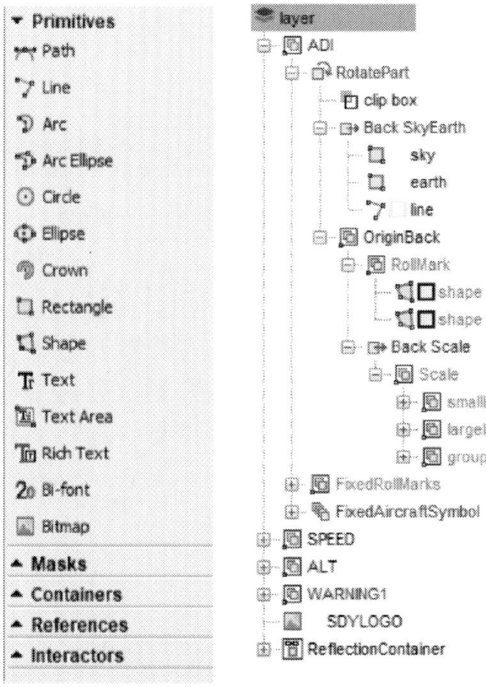

Figure 3. Hierarchy of layers (on the right) with graphical primitives like lines and groups of shapes in SCADE Display for the PFD in Figure 4. These shapes originate from the list of primitives on the icon bar on the left.

B. Limitations

Two limitations apply to this use case as it has been tested for this paper:

- the model is limited to the visual display, i.e., no functionality should be part of the model
- the model should be written in Python using an appropriate UI framework or in PGF/TikZ (i.e., LaTeX).

The latter is a limitation that results from the limitation of the LLMs type of output. Even though GPT-4 has some multimodal capabilities and can also process images as input, it cannot produce SysML models directly, but only text-based descriptions of a model. In SCADE and other tools such SysML models look like schematics in the editor of the tool, but the underlying database is saved in XML format. By this means, LLMs could output the model as an XML file, rather than outputting a graphics file representing the SysML model. For this to work, a prerequisite would be that the LLM has been trained for outputting XML and using domain-specific (tool specific) designators, naming rules and other rules to obey. It seems to be plausible that such training could be accomplished in principle. However, there are much simpler ways to connect the LLMs to the MBSE development environment: In Ansys SCADE, users can write Python code to automate all kinds of tasks using an Application Programming Interface (API). This way the LLM can hook into the development environment and generate the model directly without any file based detour. Using this API simplifies the modeling task for an AI that cannot generate images or draw schematics on its own. For the scope of this paper, even using the API would not be feasible, since using the API would afford pre-training or at least fine-tuning of the LLM. Instead, the LLMs was instructed to use Python and let it generate the graphics with the help of an appropriate framework like Qt or Tkinter that contain all visual elements (widgets) needed. As an alternative, also PGF/TikZ instructions for processing with LaTeXhave been proposed to the LLM.

The first limitation further simplifies the task for the LLMs: In order to generate a graphical model of the HMI of the PFD, no knowledge of SysML is needed. Instead, the LLM can use the API in SCADE Display to instantiate graphical primitives from a palette of basic shapes and edit their graphical properties and appearance (see Figure 3). SCADE Display also allows to implement basic functionality using logic expressions and setting properties of the shapes (e.g., visibility, color or text strings) in a way that resembles conditional statements in a programming language, but this feature cannot be known to the LLM without proper training and thus was not expected to be used by the LLM in this study.

C. Example

Figure 4 shows an example of an PFD as provided by SCADE Display for demonstration purposes. It contains some of the elements typically associated with the digital version of an artificial horizon (so-called "glass cockpit"), but there is no standardized layout and no mandatory information to be shown besides the horizon and scales for the attitude of the airplane. In those days of analog cockpits the artificial horizon did not provide information about altitude and speed and was far less cluttered with additional flight information and warning/caution indicators. Nowadays a variety of such information can be found additionally on a digital PFD.

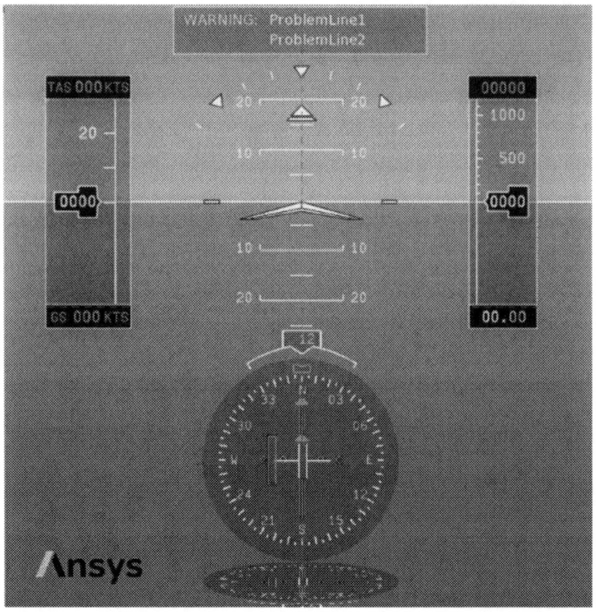

Figure 4. Graphical model of the PFD in SCADE Display as provided by the tool as an example.

In this paper, the example in Figure 4 serves as a benchmark for evaluating the degree of precision the LLMs achieve when comparing it to each generated display model. As stated in the beginning of this paper, no generally accepted quantitative measures can be used to rate the quality of the outcome. The evaluation is based on the visual assessment on how closely related the generated results are if compared to the PFD example and serves as a performance *estimate*. Generally accepted benchmarks for evaluating the performance of LLMs like HELM [10] or ARC [11] are not suited for this kind of evaluation. Even the very broad and comprehensive collection of tests and benchmarks BIG-bench [12] does not provide a method of measuring the model building capabilities from requirements as needed for this paper. In [13] the author examines over 100 benchmarks for commonsense reasoning in AI. His conclusion is that many of them are incomplete or contain flaws. As of today, there is no proven method of measuring the skills of LLMs reliably (moreover, commonsense-reasoning is different from model-building).

D. Requirements

The starting point for the LLMs is the set of requirements that specify the display of the PFD HMI as described in the previous subsection. These requirements should be made by the human engineer, as depicted in Figure 1. These requirements have been tailored to more or less match the design and structure of the PFD example in Figure 4. It represents the instrument in the supposedly simplest form and omits those types of information which may be specific to certain models of aircraft or manufacturers.

a) Full list of requirements: This list has been used for GPT-3.5 and GPT-4, but not in full length for CodeLlaMA, StarCoder and CodeGen. Only OpenAIs leading-edge products could use such a long list of requirements as *single, contiguous* input (together with the system prompt or instruction). For the other LLMs a shorted version has been used (see below).

1) General Layout & Dimensions:
   - The PFD shall have a rectangular aspect ratio suitable for installation in standard cockpit instrument panels.
   - The sky and earth shall be perfectly aligned at the horizon line.
   - The horizon line shall be centered horizontally on the PFD, and its vertical placement shall adjust based on the aircraft's pitch angle.
2) Color and Appearance:
   - The PFD shall represent the sky in blue.
   - The PFD shall represent the earth in brown.
   - The horizon line shall be a distinct, bold white line for easy visibility against both the sky and earth backdrops.
3) Aircraft Attitude Indicator:
   - An aircraft symbol, representing the relative pitch and roll of the aircraft, shall be fixed centrally on the PFD.
   - The aircraft symbol shall be displayed in a contrasting color (e.g., white) to ensure it is distinct against both sky and earth.
4) Altitude Tape:
   - The PFD shall display an altitude tape vertically on the right side, showing the current altitude of the aircraft.
   - The altitude values shall be displayed in white digits with a black outline for easy readability.
   - An arrow or pointer shall indicate the current altitude on the tape.
5) Airspeed Tape:
   - The PFD shall display an airspeed tape vertically on the left side, showing the current airspeed of the aircraft.
   - The airspeed values shall be displayed in white digits with a black outline.
   - An arrow or pointer shall indicate the current airspeed on the tape.
6) Heading Indicator:
   - The PFD shall display a horizontal heading tape or compass rose at the bottom of the display.
   - The current heading shall be indicated by a fixed pointer or triangle, with the tape/rose rotating behind it.
7) Turn Coordinator:
   - The PFD shall incorporate a turn coordinator, represented by a curved line or other suitable graphical representation, to show the rate and direction of turn.
8) Additional Flight Information:
   - The PFD shall display other pertinent flight data such as vertical speed, angle of attack, and barometric pressure.
   - This information should be arranged in a manner that does not clutter the primary attitude information.

```
9) Warning and Caution Indicators:
   • The PFD shall have provisions for
     displaying warning (red) and caution
     (amber) indications for critical flight
     parameters, such as stall warnings or
     autopilot disengagement.
```

b) Short list of requirements: The full list of requirements was too long for CodeLlaMA, StarCoder and CodeGen. CodeLlaMA did not finish the code generation process properly and stopped the output in the middle of the code - unfinished and not ready to run on the Python interpreter. StarCoder and CodeGen did not output anything, the process stopped with a time-out error. For this reason, a short list of requirements was used. This way, the modeling task was shorter also and could be finished with less tokens for the output. The short list consists of all requirements from above up to and including requirement no. 5.

E. Prompt engineering

Of course, the raw list of requirements is not enough to instruct LLMs to generate any code. Even though the requirements made up the greatest part of the input prompt, the LLMs needed to get instructions on how to code and in what language. The prompt was also used to describe the scenario and the role the LLM was expected to play when generating the code. The prompt was therefore separated into a first part which was labeled as "Instruction" and a second part which was named "Requirements". Such labeling and structuring is considered to be good practice and generally improves the outcome. Substantially better results could be expected, if instead of this "zero-shot learning" approach, at least a single example of the code to be generated would be presented to the LLM ("few-shot learning") as part of the input prompt [14]. This would have meant that a corpus of instructions was used along with an example of HMI related graphic routines. However, it was the focus of this work to only study the potential of LLMs in understanding typical (traditional) requirements in natural language and to add only a minimal amount of instructions beforehand (often referred to as "system prompt"). The following paragraph depicts the input prompt used:

```
### Instruction ###
You are a software developer who writes
code for the user interface of a Primary
Flight Display (PFD) used in an airplane's
cockpit.
Your language of choice is Python. Use
the following list of requirements as
a specification of the properties and
appearance of the user interface.
All requirements must be met. Output the
code for generating the graphics of the
user interface.
### Requirements ###
...
```

Minor changes to this input prompt were used occasionally, e.g., to instruct the LLM to use a different programming language instead of Python. For instance, GPT-4 was asked to output the code for the HMI of the primary flight display using the TikZ package of LaTeX. The advantage of this variation was, that it was perfectly clear that only static code to generate the visuals was asked for, instead of functional code that would compute changes in the aircraft's attitude from sensor inputs. In fact, the code generated by CodeLlama for the right PFD in Figure 7 comprises function calls like `getPitchAngle()`, `getRollAngle()`, `getAltitude()` and `getAirspeed()`, which are supposed to provide sensor data from real-time measurements. For the complete code, refer to listing 10 in the appendix. Such functional code was not part of the assignment and therefore the Python interpreter aborts execution after drawing the basic layout of the display, thus omitting any adjustments to be made to the indicated attitude of the aircraft or changes in altitude or speed.

No instruction finetuning was used to further improve the outcome. Chain-of-thought finetuning was also not employed, even though it should lead to substantially better results [15]–[20], given that the task of model building from requirements requires engineers to also think in a "divide and conquer" fashion and the build the system step-by-step from bottom-up. A single requirement (single sentence in natural language) in this way could be quite challenging and sophisticated and require many complex technical considerations, but be still quite feasible, if divided conceptually into sub-tasks and solved sequentially.

III. Findings

This section explains the results of a comparison of the suitability of different language models for the use case presented.

A. GPT-3.5 and GPT-4 are ahead

GPT-3.5 and GPT-4 could handle the full list of requirements, whereas the other LLMs tested in this work failed. Besides these two, only CodeLlama (CodeLlama-34b-Instruct) could at least handle the shorter list of requirements, primarily due to the limited context length of the LLM (4096 tokens for CodeLlama, see [2]).

The StarChat LLM is advertised to be the "fine-tuned versions of the StarCoder family to act as helpful coding assistants" (taken from Hugging Face website). And further: "The base model has 16B parameters and was pretrained on one trillion tokens sourced from 80+ programming languages." As StarCode offers a context window of 8K tokens [3], it was expected to actually generate some code, irrespective of the quality and the achievements. The same applies to CodeGen (codegen25-7b-instruct) from Salesforce [4]. However, running the models on Hugging Face playground led to extremely long runtimes and eventually was aborted on the server side. For this reason, the table shows 0% fulfillment rate. It remains unclear of these two LLMs would be able to process the full (or shorted) list of requirements if run on a dedicated, powerful server. Answering this question is left for future work.

In a paper titled "Sparks of Artificial General Intelligence: Early experiments with GPT-4" [21] the authors examine the capabilities of GPT-4 in graphical user interface programming. They claim that "... GPT-4 is also an expert in GUI programming, knowing how to create an accurate layout and handle complicated input events".

With this in mind, expectations were high that at least the leading-edge LLM GPT-4 could satisfactorily fulfill the task of translating requirements written in natural language into programming language code for the static display of HMI showcases. As Table I shows for the full list of requirements listed in the preceding section, GPT-4 is indeed capable of completing this task in a way that it can *assist* a human engineer in building a first, basic graphic model of a HMI for further processing with tools like SCADE Display and subsequent code generation with SCADE Display KCG or similar. In Figure 5, two of the best results are shown. The code for the PFD on the right is given in Figure 8 (code for the other on the left omitted to save space).

The metrics in Table I should be understood as meaning that the respective language model was used for several runs under the same conditions, resulting in different code variants for each run. These were then analyzed in terms of their degree of fulfillment and the dispersion characterized by the lower and upper bounds as well as the median.

The numbers indicate that in at least one case GPT-4 could successfully meet all requirements listed in Section II-D and the minimum number of requirements that could be satisfied is twice as high as in the case of GPT-3 (37% vs. 16%). The median is also almost twice as high and GPT-4 always produced code that could be run by the interpreter (Python or LaTeX) right from the start (no code fiddling needed). GPT-3.5 produced code that was erroneous in one case, but it could be corrected by the LLM itself after being instructed to do so.

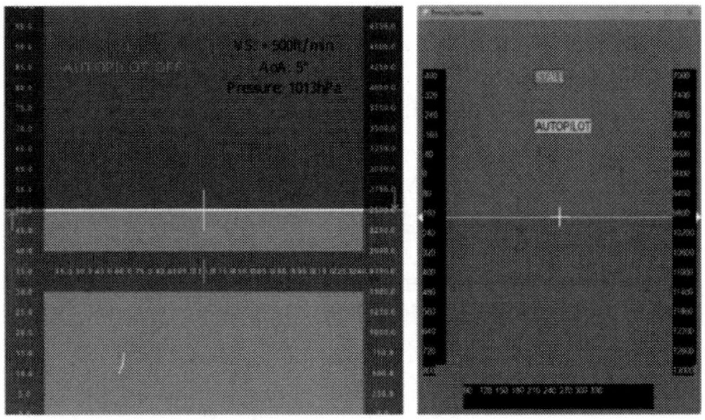

Figure 5. PFD with highest degree of fulfillment (100%, left) for the full list of requirements as generated by GPT-4 using the TikZ (LaTeX) language (see Table I). The PFD on the right side achieved 84% fulfillment.

It should be noted that the variability of the code is quite high considering all instances, despite the fact that no input prompt changes had been made and the requirements also were kept untouched. The LLMs was presented one and the same input repeatedly and the code was analyzed by comparing each and every requirement to what could actually be seen on the display of the (virtual) instrument when the code was executed. The number of satisfied requirements on the *static* display was counted and led to the percentage measure.

TABLE I
EVALUATION OF LLMS FOR THE FULL LIST OF REQUIREMENTS

	Full list of requirements (no. 1 to 9)		
	Degree of fulfillment	***# of error-free code variantes***	***# of correctable code variants***
GPT-4	Min. 37% Median 74% Max. 100%	14 of 14	N/A
GPT-3.5	Min. 16% Median 39% Max. 68%	7 of 8	1 of 8[a]
CodeLlama	0%	0 of 2[b]	0 of 2[b]
StarChat	0%	0 of 2[c]	0 of 2[c]
CodeGen2.5	0%	0 of 2[c]	0 of 2[c]

[a]contained errors that GPT-3.5 corrected after being instructed
[b]code output ended after approx. 5000 characters
[c]timeout after several minutes without any output

B. CodeLlama: Shorted list of requirements

The shorted list of requirements comprises requirement no. 1 to no. 5 and represents a very basic PFD. By this means the number of code lines for the output was essentially reduced. This enabled CodeLlama to become part of the game, i.e. it could finish the code which was otherwise aborted. Interestingly, all six runs in which CodeLlama came into operation produced code with the same type of error concerning the proper usage of the UI framework Qt. Correcting this error required the manual replacement of a line of code with three additional lines.

The complete code for the PFD on the left side of Figure 6 is given by the listing in Figure 9 in the appendix. It represents the original code from CodeLlma without the corrections. The same Qt related error produced CodeLlama in all six instances of output in Table II and could be resolved analogously.

TABLE II
EVALUATION OF LLMS FOR THE SHORT LIST OF REQUIREMENTS

	Short list of requirements (no. 1 to 5)		
	Degree of fulfillment	***# of error-free code variantes***	***# of correctable code variants***
GPT-4	100%	8 of 8	N/A
GPT-3.5	Min. 64% Median 84% Max. 96%	4 of 4	N/A
CodeLlama	Min. 14% Median 29% Max. 86%	0 of 6	6 of 6[a]
StarChat	0%	0 of 2[b]	0 of 2[b]
CodeGen2.5	0%	0 of 2[b]	0 of 2[b]

[a]repeatedly the same error using Qt, but code corrected by GPT-4
[b]timeout after several minutes without any output

In the table it can be seen that CodeLlama is far less powerful in GUI programming (i.e., HMI generation from requirements) than GPT-3.5 and certainly GPT-4, with a median degree of fulfillment of 29% vs. 84% for GPT-3.5. GPT-4 could meet all requirements and produced error-free code, as expected.

C. Visual examination and oddities

In Table II the noticeable spread between the best result from CodeLlma and the worst corresponds to the great variability that can be seen in the output of CodeLlama in Figure 6 and Figure 7. The Qt window on the left in Figure 6 is almost empty and seems to be a complete failure, but this stems from the fact that on the upper half of the window an image of the sky should be loaded from a file and in the lower half an image of the earth. The two Python instructions for loading these two files had been commented out, because they were not readily available. With those images included, the window would not look that defective and an artificial horizon would be noticeable at the boundary between the images.

Another point to mention is the aircraft symbol, which can hardly be seen in the upper center of the window. It is quite small and represented by the unicode symbol "rocket" (`U+1F680`), which was included in the source code as text string.

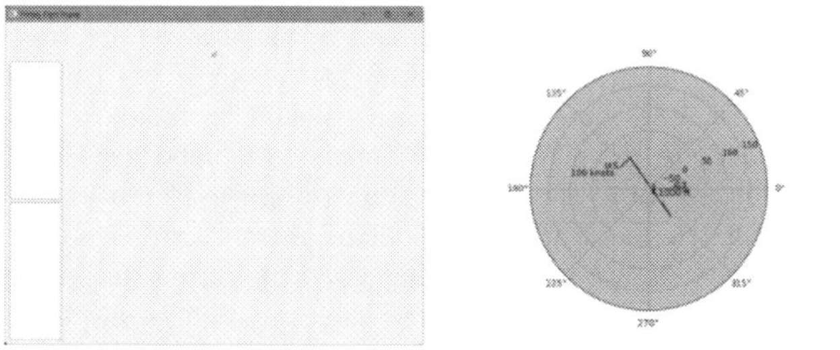

Figure 6. PFD with lowest degree of fulfillment (14%, left side) for the short list of requirements as generated by CodeLlama. CodeLlama also generated variants of the PFD on the basis of a polar diagram (29%, right side).

The variability of the results from CodeLlama is quite remarkable, as can be seen on the right side of Figure 6. In this case the LLM tried to use a polar diagram to fit the PFD in, but with this approach it sacrificed many requirements so that only 29% could be met.

In those cases in which CodeLlama "decided" to chose the right UI/GUI framework and a suitable graph paradigm, the outcome was not so bad as can be seen in Figure 7. On one hand, the aircraft symbol was merely a circle in the middle of the display, but the requirements did not specify how it should look like well enough on the other hand. It can be stated that *what is not specified thoroughly, precisely and comprehensively in the requirements can be implemented by the LLM modeling assistant freely and with little common-sense knowledge and engineering experience, it will be.* The more common-sense knowledge an advanced LLM has, the better it can fill those gaps in the specification and thereby interpret the human will and serve the intended purpose.

For this reason, CodeLlama generated the "rubber bands" (airspeed tape and altitude tape, requirements no. 4 and 5) on the left and right side in such a way that it is hard to use from a practical standpoint, but it also fulfilled the needs written down in the requirements. It was just lacking the knowledge, that a cluttered display with multiple symbols and text snippets overlaying each other is basically useless.

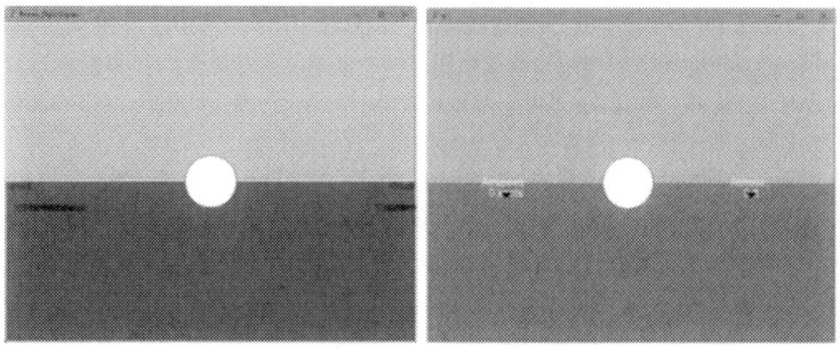

Figure 7. PFD variants with highest degree of fulfillment (57% and 86%) for the short list of requirements as generated by CodeLlama. Note that GPT-3.5 and GPT-4 were still much better (see Table II).

GPT-4, however, seems to be much more aware of the implications of certain design and layout related decisions it makes for the practical usability, as can be clearly seen in Figure 5. It knows that the altitude is typically displayed in quantaties of 100 feet and the speed in a finer resolution (here 5 knots). The compass rose (heading tape) in PFD on the right correctly shows values of degree ranging from 90 to 330, but not surpassing 360 degree. All of this was not specified by the requirements.

Interestingly, GPT-4 did not realize that it would be better to let the altitude increase towards the sky and sketched the numbers the other way around (higher altitude towards the earth). A human engineer would have done it opposed for sure, even without a corresponding requirement. Such implicit attributes must be derived from the requirements as part of the engineering task. This is something that GPT-4 is not able to do in each and every case. Its common-sense knowledge is incomplete, otherwise GPT-4 would not provide numbers for negative speeds on the left airspeed tape. Numbers can be negative in many cases, but for speed it makes no sense. It is speculative, but maybe GPT-4 would not have included negative values if the tape and its purpose was titled "speed" instead of "*air*speed".

Taking the speculation one step further, a single example of a PFD with negative speed values can be found on the internet if a search (using Google) for images for "primary flight display" is performed. The image with negative speed values leads to GitHub: In 2019, under the name "kouky" an author published a project for a PFD to be used in micro UAVs (Unmanned Aerial Vehicles) [22]. Such aerial vehicles or multicopters can indeed exhibit negative speeds if flown backwards. If GPT-4 included the negative speeds intentionally, the LLM either learned from the code of this project on GitHub or from the corresponding image.

IV. Conclusion and Future Work

The following section examines the steps that need to be taken to advance the concept presented and summarizes the findings of this study.

A. Next Steps

The work presented in this paper will be the starting point for building a modelling assistant for HMI generation from requirements based on open source LLMs like Llama 2 (i.e. CodeLlama), StarCoder (StarChat) or comparable LLMs, provided that these can be fine-tuned to achieve similar results as GPT-4. In [23] and [24], the authors give practical hints and instructions on how to fine-tune OpenLlama and StarCoder, respectively. It is an open question, whether or not these open LLMs can be fine-tuned in that way or training from scratch is needed. The fact that StarChat (and CodeGen) could not be included in the comparative study in this paper in a proper way due to the long runtimes and the timeout on HuggingFace does *not* imply that these LLMs should be not suited at all. On specialized hardware with enough capacity for running the LLMs exclusively, it should be possible to actually produce an outcome.

The finetuning task should in the intermediate run also include the training on the proper usage of the Python API of the SCADE MBSE framework (Model Based Systems Engineering) or similar development platforms (IBM Rational Rhapsody or Mathworks MATLAB/Simulink, depending on API suitability). This way the LLMs could produce the code for the HMI model directly. In SCADE Display this model would then be revised by the human engineer as depicted in Figure 1, before the code generator KCG would generate the actual C or Ada code. In the case of SCADE Display KCG, the tool uses the OpenGL SC subset of the graphic library explicitly targeted at safety critical applications ("SC" standing for "safety critical").

In the long run the whole concept should be rolled out on model building in general, i.e., SysML based models of functional components, not only HMI/GUI/UI use cases. This will be the hardest part and requires a deep understanding of the system or subsystem to be developed. In in the scope of this work it was not possible to show the feasibility of such a modeling task. Presumably, it is a very long way from graphical model models for displaying purposes to functional models in a broad sense. Future work should examine the chances that such AI based assistants could support tomorrow's engineers in the development process for software systems in general. Safety critical applications will not impose barriers if the AI assistant comes into play in the right phase of the process, as suggested by this paper.

B. Summary

In this paper it was shown that an AI/LLM assisted software development process without the need for manual coding is possible, if - instead of generating the final C/Ada source code - the LLM is instructed to create a *model* of the system. The term "system" in this respect refers to a specific system for displaying information and human interaction as in HMI applications or for GUI/UI use cases. No modeling capabilities for functional components were included in this study, primarily due to the lack of an appropriate output format for the model itself (XML/SysML). In this scenario, the LLM was given a collection of requirements for the *visual* component of a software system and then instructed to translate these requirements into a graphical model of the HMI to be displayed, including basic colored shapes (rectangles, circles, etc.), text insets, call-outs, etc. - all arranged and adjusted to the fulfill the requirements. The results five from different LLMs were studied. However, only three of the LLMs produced comparable results (due to limitations concerning the computing platform).

Comparing the results, it was shown that GPT-4 is superior in performance and accuracy. The outcome in general shows a great amount of variability including visual forms of confabulation if details are left out or specified in an unprecise manner. GPT-4 does not produce contradictory objects with respect to the requirements, but also needs more self-explanatory, technically explicit guidelines than the requirements typically used by today's engineers provide. The LLMs have deficiencies in common-sense reasoning and fill gaps in background knowledge either by figures stemming from misleading training data or by other unknown influencing factors. For this reason, a set of requirements with a fine granularity is important, i.e., the level of detail is crucial. However, the number of requirements that can be processed by the LLM in one single run is very limited. Because of this, techniques like Chain-Of-Thought (COT) prompt enigneering should be employed in the instructions in which the requirements are embedded. LLMs benefit from a large context window (tokens to be processed in a single run) and the capabilities to process long documents (the requirements) must be improved by current techniques and future enhancements of the language models in order to fulfill the expectations and the practical usability of this whole concept.

It has been argued that the usage of LLM assistants for a no-code software development process is not prohibitive even for safety critical fields of application like cockpit instruments in aircrafts. The phase in which the LLM is employed and what task it is instructed to perform (i.e., model building instead of source code generation) is crucial and the human engineers always needs to stay in the loop by checking and revising the models. The use case studied was a Primary Flight Display (PFD) as used by pilots and served as an indicative measure for the performance of selected LLMs with coding capabilities. The study was performed on an empirical basis on this single use case, so no universal validity for other scenarios is claimed.

References

[1] OpenAI, "Gpt-4 technical report," *ArXiv*, vol. abs/2303.08774, 2023. [Online]. Available from: https://arxiv.org/abs/2303.08774 [retrieved: Oct., 2023]

[2] H. Touvron *et al.*, "Llama 2: Open foundation and fine-tuned chat models," 7 2023. [Online]. Available from: https://www.semanticscholar.org/paper/104b0bb1da562d53cbda87aec79ef6a2827d191a [retrieved: Oct., 2023]

[3] R. Li *et al.*, "Starcoder: may the source be with you!" 2023. [Online]. Available from: http://arxiv.org/abs/2305.06161 [retrieved: Oct., 2023]

[4] E. Nijkamp, H. Hayashi, C. Xiong, S. Savarese, and Y. Zhou, "Codegen2: Lessons for training llms on programming and natural languages," *arXiv preprint*, 2023.

[5] R. Bagnara, A. Bagnara, and P. M. Hill, "The misra c coding standard and its role in the development and analysis of safety- and security-critical embedded software." *CoRR*, vol. abs/1809.00821, 2018. [Online]. Available from: http://dblp.uni-trier.de/db/journals/corr/corr1809.html#abs-1809-00821 [retrieved: Oct., 2023]

[6] R. Bagnara, M. Barr, and P. M. Hill, "Barr-c: 2018 and misra c: 2012: Synergy between the two most widely used c coding standards." *CoRR*, vol. abs/2003.06893, 2020. [Online]. Available from: http://dblp.uni-trier.de/db/journals/corr/corr2003.html#abs-2003-06893 [retrieved: Oct., 2023]

[7] J. Holt, *SysML for Systems Engineering: A Model-Based Approach*, ser. Computing. Institution of Engineering and Technology, 2018. [Online]. Available from: https://digital-library.theiet.org/content/books/pc/pbpc020e [retrieved: Oct., 2023]

[8] D. Iqbal, A. Abbas, M. Ali, M. U. S. Khan, and R. Nawaz, "Requirement validation for embedded systems in automotive industry through modeling," *IEEE Access*, vol. PP, pp. 1–1, 01 2020.

[9] H. Touvron *et al.*, "Llama: Open and efficient foundation language models," 2023. [Online]. Available from: http://arxiv.org/abs/2302.13971 [retrieved: Oct., 2023]

[10] P. Liang *et al.*, "Holistic evaluation of language models," 2022. [Online]. Available from: https://arxiv.org/abs/2211.09110 [retrieved: Oct., 2023]

[11] F. Chollet, "On the measure of intelligence," 2019. [Online]. Available from: http://arxiv.org/abs/1911.01547 [retrieved: Oct., 2023]

[12] A. Srivastava *et al.*, "Beyond the imitation game: Quantifying and extrapolating the capabilities of language models," 2023. [Online]. Available from: http://arxiv.org/abs/2206.04615 [retrieved: Oct., 2023]

[13] E. Davis, "Benchmarks for automated commonsense reasoning: A survey," 2023. [Online]. Available from: https://arxiv.org/abs/2302.04752 [retrieved: Oct., 2023]

[14] T. B. Brown *et al.*, "Language models are few-shot learners," 2020. [Online]. Available from: https://arxiv.org/abs/2005.14165 [retrieved: Oct., 2023]

[15] H. W. Chung *et al.*, "Scaling instruction-finetuned language models," 2022. [Online]. Available from: https://arxiv.org/abs/2210.11416 [retrieved: Oct., 2023]

[16] T. Z. Zhao, E. Wallace, S. Feng, D. Klein, and S. Singh, "Calibrate before use: Improving few-shot performance of language models," 2021. [Online]. Available from: https://arxiv.org/abs/2102.09690 [retrieved: Oct., 2023]

[17] H. Su *et al.*, "Selective annotation makes language models better few-shot learners," 2022. [Online]. Available from: https://arxiv.org/abs/2209.01975 [retrieved: Oct., 2023]

[18] S. Diao, P. Wang, Y. Lin, and T. Zhang, "Active prompting with chain-of-thought for large language models," 2023. [Online]. Available from: https://arxiv.org/abs/2302.12246 [retrieved: Oct., 2023]

[19] M. Chen *et al.*, "Evaluating large language models trained on code," 2021. [Online]. Available from: https://arxiv.org/abs/2107.03374 [retrieved: Oct., 2023]

[20] L. Ouyang *et al.*, "Training language models to follow instructions with human feedback," 2022. [Online]. Available from: https://arxiv.org/abs/2203.02155 [retrieved: Oct., 2023]

[21] S. Bubeck *et al.*, "Sparks of artificial general intelligence: Early experiments with gpt-4," 2023. [Online]. Available from: https://arxiv.org/abs/2303.12712 [retrieved: Oct., 2023]

[22] kouky (nickname), "Primary flight display graphics framework for mac + ios," 2022. [Online]. Available from: https://github.com/kouky/PrimaryFlightDisplay [retrieved: Oct., 2023]

[23] M. Thissen, "Feintuning großer sprachmodelle – so gehts," *iX*, vol. 10, pp. 50–59, 2023.

[24] R. Wartala, "Eigene coding-assistenten trainieren und nutzen," *iX*, vol. 10, pp. 44–49, 2023.

APPENDIX

C. Source code examples

GPT-4

```
import pygame
import sys

# Initialize pygame
pygame.init()

# Set dimensions
WIDTH, HEIGHT = 800, 800

# Colors
BLUE = (0, 102, 204)
BROWN = (153, 76, 0)
WHITE = (255, 255, 255)
BLACK = (0, 0, 0)
RED = (255, 0, 0)
AMBER = (255, 191, 0)

# Create screen and clock
screen = pygame.display.set_mode((WIDTH, HEIGHT))
pygame.display.set_caption("Primary Flight Display")
clock = pygame.time.Clock()

def draw_horizon(pitch_angle=0):
    pygame.draw.rect(screen, BLUE, (0, 0, WIDTH, HEIGHT // 2 -
    pitch_angle))
    pygame.draw.rect(screen, BROWN, (0, HEIGHT // 2 - pitch_angle,
    WIDTH, HEIGHT))
    pygame.draw.line(screen, WHITE, (0, HEIGHT // 2 - pitch_angle),
    (WIDTH, HEIGHT // 2 - pitch_angle), 2)

def draw_aircraft_symbol():
    pygame.draw.line(screen, WHITE, (WIDTH//2 - 20, HEIGHT//2),
    (WIDTH//2 + 20, HEIGHT//2), 3)
    pygame.draw.line(screen, WHITE, (WIDTH//2, HEIGHT//2 - 20),
    (WIDTH//2, HEIGHT//2 + 20), 3)

def draw_altitude_tape(altitude=10000):
    start_y = 100
    end_y = 700
    pygame.draw.rect(screen, BLACK, (WIDTH - 60, start_y, 50, end_y
    - start_y))
    pygame.draw.polygon(screen, WHITE, [(WIDTH-10, HEIGHT//2-10),
    (WIDTH-10, HEIGHT//2+10), (WIDTH, HEIGHT//2)])
    for y in range(start_y, end_y + 1, 40):
        altitude_value = altitude - (HEIGHT//2 - y) * 10
        text_surface = pygame.font.SysFont('arial',
        20).render(str(altitude_value), True, WHITE, BLACK)
        screen.blit(text_surface, (WIDTH - 60, y))

def draw_airspeed_tape(speed=200):
    start_y = 100
    end_y = 700
    pygame.draw.rect(screen, BLACK, (10, start_y, 50, end_y -
    start_y))
    pygame.draw.polygon(screen, WHITE, [(10, HEIGHT//2-10), (10,
    HEIGHT//2+10), (0, HEIGHT//2)])
    for y in range(start_y, end_y + 1, 40):
        speed_value = speed - (HEIGHT//2 - y) * 2
        text_surface = pygame.font.SysFont('arial',
        20).render(str(speed_value), True, WHITE, BLACK)
        screen.blit(text_surface, (10, y))

def draw_heading_indicator(heading=90):
    pygame.draw.rect(screen, BLACK, (100, HEIGHT - 60, WIDTH - 200,
    50))
    for i in range(0, 360, 30):
        x_offset = (i - heading) * (WIDTH-200) / 360
        if 100 + x_offset > 90 and 100 + x_offset < WIDTH - 90:
            text_surface = pygame.font.SysFont('arial',
            20).render(str(i), True, WHITE, BLACK)
            screen.blit(text_surface, (100 + x_offset, HEIGHT -
            60))

def display_warning(message):
    text_surface = pygame.font.SysFont('arial', 25).render(message,
    True, WHITE, RED)
    screen.blit(text_surface, (WIDTH // 2 - 50, HEIGHT // 8))

def display_caution(message):
    text_surface = pygame.font.SysFont('arial', 25).render(message,
    True, BLACK, AMBER)
    screen.blit(text_surface, (WIDTH // 2 - 50, 2 * HEIGHT // 8))

# Main loop
while True:
    for event in pygame.event.get():
        if event.type == pygame.QUIT:
            pygame.quit()
            sys.exit()

    screen.fill(BLACK)

    draw_horizon()
    draw_aircraft_symbol()
    draw_altitude_tape()
    draw_airspeed_tape()
    draw_heading_indicator()

    display_warning("STALL")
    display_caution("AUTOPILOT")

    pygame.display.flip()
    clock.tick(60)
```

Figure 8. Python code generated by GPT-4 for the full list of requirements

CodeLlama-34b-Instruct

```
import matplotlib.pyplot as plt
from PyQt5 import QtWidgets

class PrimaryFlightDisplay(QtWidgets.QMainWindow):
    def __init__(self):
        super().__init__()

        # Set up the main window
        self.setGeometry(0, 0, 800, 600)
        self.setWindowTitle("Primary Flight Display")

        # Create the layout
        self.layout = QtWidgets.QVBoxLayout()
        self.setCentralWidget(self.layout)

        # Add the sky and earth images
        self.sky_image = plt.imread('sky.png')
        self.earth_image = plt.imread('earth.png')
        self.sky_widget = QtWidgets.QLabel()
        self.earth_widget = QtWidgets.QLabel()
        self.sky_widget.setPixmap(self.sky_image)
        self.earth_widget.setPixmap(self.earth_image)
        self.layout.addWidget(self.sky_widget)
        self.layout.addWidget(self.earth_widget)

        # Add the horizon line
        self.horizon_line = QtWidgets.QFrame()
        self.horizon_line.setStyleSheet("background-color: white;
border: none; height: 2px;")
        self.layout.addWidget(self.horizon_line)

        # Add the aircraft attitude indicator
        self.aircraft_symbol = QtWidgets.QLabel()
        self.aircraft_symbol.setText("✈")
        self.aircraft_symbol.setAlignment(QtCore.Qt.AlignCenter)
        self.aircraft_symbol.setFontSize(24)
        self.layout.addWidget(self.aircraft_symbol)

        # Add the altitude tape
        self.altitude_tape = QtWidgets.QScrollArea()
        self.altitude_tape.setVerticalScrollBarPolicy(
                QtCore.Qt.ScrollBarAsNeeded)
        self.altitude_tape.setHorizontalScrollBarPolicy(
                QtCore.Qt.ScrollBarAlwaysOff)
        self.altitude_tape.setWidgetResizable(True)
        self.altitude_tape.setFixedWidth(100)
        self.altitude_tape.setMinimumHeight(100)
        self.altitude_tape.setMaximumHeight(300)
        self.altitude_tape.setBackgroundRole(QtGui.QPalette.Base)
        self.altitude_tape.setAutoFillBackground(True)
        self.altitude_tape.show()
        self.layout.addWidget(self.altitude_tape)

        # Add the airspeed tape
        self.airspeed_tape = QtWidgets.QScrollArea()
        self.airspeed_tape.setVerticalScrollBarPolicy(
                QtCore.Qt.ScrollBarAsNeeded)
        self.airspeed_tape.setHorizontalScrollBarPolicy(
                QtCore.Qt.ScrollBarAlwaysOff)
        self.airspeed_tape.setWidgetResizable(True)
        self.airspeed_tape.setFixedWidth(100)
        self.airspeed_tape.setMinimumHeight(100)
        self.airspeed_tape.setMaximumHeight(300)
        self.airspeed_tape.setBackgroundRole(QtGui.QPalette.Base)
        self.airspeed_tape.setAutoFillBackground(True)
        self.airspeed_tape.show()
        self.layout.addWidget(self.airspeed_tape)

    def update_display(self, aircraft_pitch, aircraft_roll,
    altitude, airspeed):
        # Update the aircraft attitude indicator
        self.aircraft_symbol.move(self.width() / 2 -
        self.aircraft_symbol.width() / 2, self.height() / 2 -
        self.aircraft_symbol.height() / 2 + aircraft_pitch * 10)
        self.aircraft_symbol.rotate(-aircraft_roll)

        # Update the altitude tape
        self.altitude_tape.verticalScrollBar().setValue(altitude)

        # Update the airspeed tape
        self.airspeed_tape.verticalScrollBar().setValue(airspeed)

if __name__ == "__main__":
    app = QtWidgets.QApplication([])
    pfd = PrimaryFlightDisplay()
    pfd.show()
```

Figure 9. Python code generated by CodeLlama for the short list of requirements for the HMI of a PFD. Note that it contained flaws (wrong usage of Qt and missing image files) that could be corrected manually. The (corrected) code refers to the left side in Figure 6.

CodeLlama-34b-Instruct

```
import tkinter as tk
from math import sin, cos, radians

# Define constants for the dimensions and colors of the PFD
WIDTH = 800
HEIGHT = 600
SKY_COLOR = "#7ec0ee" # light blue
EARTH_COLOR = "#9c6f4a" # dark brown
HORIZON_LINE_COLOR = "white"
ALTITUDE_TAPE_COLOR = "white"
AIRSPEED_TAPE_COLOR = "white"
ARROW_COLOR = "black"

# Create a canvas object to draw on
canvas = tk.Canvas(width=WIDTH, height=HEIGHT, bg="gray")
canvas.pack()

# Draw the sky and earth
sky = canvas.create_rectangle(0, 0, WIDTH, HEIGHT//2,
fill=SKY_COLOR)
earth = canvas.create_rectangle(0, HEIGHT//2, WIDTH, HEIGHT,
fill=EARTH_COLOR)

# Draw the horizon line
horizon_line = canvas.create_line(0, HEIGHT//2, WIDTH, HEIGHT//2,
fill=HORIZON_LINE_COLOR)

# Draw the aircraft attitude indicator
aircraft_symbol = canvas.create_oval(WIDTH//2-50, HEIGHT//2-50,
WIDTH//2+50, HEIGHT//2+50, fill="#fff", outline="#000")

# Draw the altitude tape
altitude_tape = canvas.create_text(WIDTH*0.8, HEIGHT*0.5,
text="Altitude: ", font=("Helvetica", 16),
fill=ALTITUDE_TAPE_COLOR)
altitude_value = canvas.create_text(WIDTH*0.8, HEIGHT*0.5 + 20,
text="0 ft", font=("Helvetica", 16), fill=ALTITUDE_TAPE_COLOR)
arrow = canvas.create_polygon((WIDTH*0.8 - 10, HEIGHT*0.5 + 20),
(WIDTH*0.8 + 10, HEIGHT*0.5 + 20), (WIDTH*0.8, HEIGHT*0.5 + 30),
fill=ARROW_COLOR)

# Draw the airspeed tape
airspeed_tape = canvas.create_text(WIDTH*0.2, HEIGHT*0.5,
text="Airspeed: ", font=("Helvetica", 16),
fill=AIRSPEED_TAPE_COLOR)
airspeed_value = canvas.create_text(WIDTH*0.2, HEIGHT*0.5 + 20,
text="0 knots", font=("Helvetica", 16), fill=AIRSPEED_TAPE_COLOR)
arrow = canvas.create_polygon((WIDTH*0.2 - 10, HEIGHT*0.5 + 20),
(WIDTH*0.2 + 10, HEIGHT*0.5 + 20), (WIDTH*0.2, HEIGHT*0.5 + 30),
fill=ARROW_COLOR)

def update():
    # Update the position of the aircraft symbol based on the
    aircraft's pitch and roll angles
    pitch = getPitchAngle() * radians(180)/pi
    roll = getRollAngle() * radians(180)/pi
    x = WIDTH//2 + 50 * sin(pitch)
    y = HEIGHT//2 + 50 * cos(roll)
    aircraft_symbol.coords(x, y)

    # Update the altitude value on the altitude tape
    altitude = getAltitude()
    altitude_value.set_text("%d feet" % altitude)
    arrow.moveTo(WIDTH*0.8, HEIGHT*0.5 + 20)

    # Update the airspeed value on the airspeed tape
    airspeed = getAirspeed()
    airspeed_value.set_text("%d knots" % airspeed)
    arrow.moveTo(WIDTH*0.2, HEIGHT*0.5 + 20)

root.after(1000, update)
```

Figure 10. Variation of the Python code generated by CodeLlama for the short list of requirements. The code refers to the right side in Figure 7 and gets aborted by the Python interpreter after initial drawing of the basic layout due to missing sensor data.

Using ChatGPT-4 for the Identification of Common UX Factors within a Pool of Measurement Items from Established UX Questionnaires

Stefan Graser, Stephan Böhm
CAEBUS Center of Advanced E-Business Studies
RheinMain University of Applied Sciences
Wiesbaden, Germany
e-mail: {stefan.graser, stephan.boehm}@hs-rm.de

Martin Schrepp
SAP SE
Walldorf, Germany
e-mail: martin.schrepp@sap.com

***Abstract*—Measuring User Experience (UX) with standardized questionnaires is a widely used method. A questionnaire is based on different scales that represent UX factors and items. However, the questionnaires have no common ground concerning naming different factors and the items used to measure them. This study aims to identify general UX factors based on the formulation of the measurement items. Items from a set of 40 established UX questionnaires were analyzed by Generative AI (GenAI) to identify semantically similar items and to cluster similar topics. We used the LLM ChatGPT-4 for this analysis. Results show that ChatGPT-4 can classify items into meaningful topics and thus help to create a deeper understanding of the structure of the UX research field. In addition, we show that ChatGPT-4 can filter items related to a predefined UX concept out of a pool of UX items.**

Keywords–User Experience (UX); UX Measurement; UX Factors; Measurement Items; Generative AI (GenAI); Large Language Model (LLM); ChatGPT; Semantic Textual Similarity (STS).

I. Introduction

User Experience (UX) is a holistic concept in Human-Computer-Interaction (HCI) describing the perception towards the use and interaction of a product, service, or system [1]. A positive UX is essential for interacting with products and services. This user's perception must be considered to gather insights into improving the UX [2]. Therefore, various methods can be found for UX measurement. The most common way to measure the UX is through standardized questionnaires providing self-reported data by the user [3]. These questionnaires can be applied in a cost-efficient, simple, and fast way [3][4].

Over the last decades, different standardized questionnaires were developed, breaking down and measuring the construct of UX. Therefore, the questionnaires refer to a holistic view or focus on a specific dimension. To be more precise, a questionnaire is based on the different factors, items, and scales about the respective dimension [5][6]. However, there is no common ground within the factors and items among the standardized UX questionnaires. Differently named factors can measure the same, but factors with the same name can measure something different [7]. This leads to a blurring of the respective measurement focus among the questionnaires. Nevertheless, a clear distinction between the measurement items is necessary to measure the same and have a shared understanding of the construct of UX. There is a lack of sufficient exposition of what different developed scales semantically mean [7].

In this regard, this study focuses on the level of the different items describing the UX dimensions. We aim to identify semantically similar items by applying Generative AI. Therefore, we used ChatGPT-4 as a Large Language Model (LLM) to analyze and compare items concerning their Semantic Textual Similarity (STS). Based on this, similar items were clustered. As a result, we try to identify UX topics from these clusters. Against this background, we address the following research questions:

RQ1: *Is Generative AI able to identify useful similarity topics based on measurement items?*

RQ2: *Which topics based on semantically similar measurement items can be identified among the most established UX questionnaires?*

This article is structured as follows: Section 2 describes the theoretical foundation of this approach. Section 3 shows related work concerning the consolidation of UX factors and common ground in UX research. Section 4 illustrates the methodological approach by applying the LLM ChatGPT-4 as Generative AI. Results are shown in Section 5. A conclusion and outlook is given in Section 6.

II. Theoretical Foundation

A. Concept of UX

As already described, UX is a multidimensional construct consisting of different dimensions and quality aspects. Usability, which is defined as "the extent to which a product can be used by specified users to achieve specified goals with effectiveness, efficiency, and satisfaction in a specified context of use" [1] is focused on completing tasks and achieving goals. UX, on the other hand, encompasses a broader spectrum of qualities related to a product's subjective impression. This includes, for example, aspects such as aesthetics or fun of use. Thus, usability can be declared a subset of UX [8].

Based on this, Hassenzahl [9] presents a distinction between pragmatic and hedonic properties. Pragmatic qualities are task-related, whereas hedonic qualities refer to non-task-related qualities [9]. However, this distinction is accompanied by problems. Firstly, a clear distinction is not always possible for a specific product. Secondly, pragmatic qualities relate to a common concept as they are task-related whereas hedonic qualities do not follow such a concept [6].

Schrepp et al. [6] followed a new approach conceptualizing UX as a defined set of quality aspects. A "UX quality aspect describes the subjective impression of users towards a semantically clearly described aspect of product usage or product design" [6]. This results in clearly described and distinct aspects that can be used to evaluate the subjective experience towards a product [6].

B. Semantic and Empirical Similarity

In this paper, we focus on investigating the semantic similarity of measurement items from UX questionnaires. Semantic similarity refers to the degree of likeness or resemblance between the item texts based on their meaning. Thus, semantic similarity expresses how closely related the underlying textual concepts are, rather than just the surface-level syntactic or structural similarity. Semantic similarity takes into account the context, relationships, and associations between words or phrases to determine their level of similarity [10]–[12]. Different statistics-based methods in Natural Language Processing (NLP) for Semantic Textual Similarity measurement can be found in the literature [10][13]–[20]. In general, the methods can be divided into the three categories Matrix Based Methods, Word Distance-Based Methods, and Sentence Embedding Based Methods [21].

Large Language Models, like GPT, use word embeddings (dense vector representations of words derived with the help of deep learning mechanisms applied to vast volumes of existing texts) to calculate semantic similarity. Thus, they are obviously helpful tools for analyzing the semantic similarity of UX items.

However, in interpreting the results of such an analysis of semantic item similarity, we must distinguish the semantic similarity of items from their empirical similarity [22][23], i.e., their empirical correlation, to understand the benefits and limitations of such an approach. We may observe items that have a small semantic similarity as estimated by an LLM but show in empirical studies quite substantial correlations.

A well-investigated example is the observation that beautiful products are perceived as usable [24][25]. Thus, visual aesthetics influence the perception of classical UX aspects like *Efficiency*, *Learnability*, or *Controllability*, and items measuring these semantically quite different aspects correlate. A similar effect exists also in the opposite direction, i.e., the perception of *Usability* influences the perception of beauty [26][27].

There are several explanations (which in fact may all contribute to the effect) for such first-sight strange empirical dependencies, for example, the general impression model [28], evaluative consistency [29], or mediator effects [30]. Another explanation is that aesthetics and usability share, in fact, some common aspects. Balance, symmetry, and order [31] or alignment [32] influence the aesthetic impression. But a UI that looks clean, ordered, and properly aligned is also easy to scan and thus, users can find elements faster and orient more easily on such an interface. Hence, it will also benefit *Efficiency* or *Learnability* [23].

Given these arguments, we can expect that items with a high semantic similarity will also show empirically high correlations (they ask for highly similar UX aspects thus, participants of a survey should give highly similar answers). However, there may be items with quite low semantic similarities but quite high empirical correlations due to the effects described above. Thus, we should not expect that we can reconstruct typical scales of established questionnaires by a purely semantical analysis of the items. Such scales are usually developed by an empirical process of item reduction, mostly by main component analysis and group items based on empirical correlations from larger studies.

C. UX Questionnaires

Quantitative UX evaluation is usually based on questionnaires as subjective assessments of user's perceptions. Various standardized UX questionnaires can be found in scientific literature. For example, Schrepp [7] describes 40 quite common UX questionnaires [7]. Every questionnaire is based on specific factors, items, and scales. Moreover, measurement focus can differ among the questionnaires. The selection of the specific questionnaire may differ depending on the application purpose or objective of the investigation.

Díaz-Oreiro et al. [33] investigated the User Experience Questionnaire UEQ [34] as the most widely used questionnaire for UX evaluation. This can be confirmed by further research [33]. The UEQ developed by Laugwitz et al. [34] is based on the UX framework by Hassenzahl [9][34]. The questionnaire consists of six factors divided into pragmatic and hedonic properties. Each factor contains four items formulated as a semantic differential scale measured by a 7-point Likert scale. The factors with their descriptions are shown below:

- **Attractiveness**: Overall impression of the product. Do users like or dislike it?
- **Perspicuity**: Is it easy to get familiar with the product and to learn how to use it?
- **Efficiency**: Can users solve their tasks without unnecessary effort? Does it react fast?
- **Dependability**: Does the user feel in control of the interaction? Is it secure and predictable?
- **Stimulation**: Is it exciting and motivating to use the product? Is it fun to use?
- **Novelty**: Is the design of the product creative? Does it catch the interest of users?

The questionnaire aims to gather a holistic impression referring to the UX of interactive products. The UEQ is an example of a questionnaire with scales representing quite abstract UX concepts and can thus be applied to many different products. The items are semantic differentials, i.e., pairs of terms with opposite meanings that represent a semantic scale (for example, slow/fast). Further details can be found online [35].

Other established questionnaires follow a different measurement concept in that their items and scales refer to concrete interface elements. For example, the Purdue Usability Testing Questionnaire [36] contains items like "Is the cursor placement consistent?" or "Does it provide visually distinctive data fields?". This form of items is much more concrete but can only be applied to a certain type of product. In addition, there are several questionnaires that can be applied only for special application domains, for example, web pages, e-commerce, or games (for an overview of common questionnaires and item formulations, see [37]). This huge variety in the way items are formulated makes it also quite challenging to categorize them concerning their semantic meaning.

No questionnaire can cover all UX factors. As already described, each questionnaire refers to a specific focus. Therefore, it is a common way to combine or apply several questionnaires simultaneously to cover all relevant aspects. Due to different items and scales, it may be more difficult for participants to complete the evaluation. Therefore, Schrepp and Thomaschewski (2019) developed the UEQ+, a modular framework. The framework is based on described factors with their respective items covering the construct UX as broadly as possible. Researchers can choose from a set of 16 UX quality aspects according to the respective product to evaluate and create an individualized UX questionnaire [38]. Further information can be found online [39].

III. Research Objective and Related Work

Due to the high number of UX questionnaires developed in the last decades, many different factors and items can be found. This emphasizes the lack of common ground within quantitative UX evaluation. Concerning this research gap, only a little research was done to consolidate general UX factors and find a common understanding.

[40] aimed to consolidate a list of general UX factors. Therefore, existing questionnaires and literature were analyzed. All collected factors were then consolidated based on their definition. This resulted in a consolidated list of general UX factors [40]. The same approach was conducted by [5] and [6]. The latest list of consolidated UX factors is shown in the following table (see Table I):

TABLE I: CONSOLIDATED UX FACTORS BASED ON [6].

(#)	Factor	Descriptive Question
(1)	Perspicuity	Is it easy to get familiar with the product and to learn how to use it?
(2)	Efficiency	Can users solve their tasks without unnecessary effort? Does the product react fast?
(3)	Dependability	Does the user feel in control of the interaction? Does the product react predictably and consistently to user commands?
(4)	Usefulness	Does using the product bring advantages to the user? Does using the product save time and effort?
(5)	Intuitive use	Can the product be used immediately without any training or help?
(6)	Adaptability	Can the product be adapted to personal preferences or personal working styles
(7)	Novelty	Is the design of the product creative? Does it catch the interest of users?
(8)	Stimulation	Is it exciting and motivating to use the product? Is it fun to use?
(9)	Clarity	Does the user interface of the product look ordered, tidy, and clear?
(10)	Quality of Content	Is the information provided by the product always actual and of good quality
(11)	Immersion	Does the user forget time and sink completely into the interaction with the product
(12)	Aesthetics	Does the product look beautiful and appealing?
(13)	Identity	Does the product help the user to socialize and to present themselves positively to other people?
(14)	Loyalty	Do people stick with the product even if there are alternative products for the same task
(15)	Trust	Do users think that their data is in safe hands and not misused to harm them?
(16)	Value	Does the product design look professional and of high quality?

Typically, UX factors are constructed with the help of empirical methods of item reduction, for example, main component analysis. Thus, items are grouped into factors based on their empirical correlations. This leads sometimes to scales that consist of items that represent, at least at first sight, semantically different concepts. Thus, it is sometimes difficult to clearly describe what the semantic behind a scale actually is. To get a deeper understanding of the concept of UX, it makes thus sense to analyze the purely semantic similarities of items and to investigate a structuring based on this concept.

Only two studies have yet applied methods to measure semantic textual similarity in the field of UX research concerning UX measurement items. Both studies applied NLP techniques at the level of the measurement items. In particular, the semantic textual similarity between the measurement items was analyzed. By doing this, the researchers tried to ensure a more accurate distinction. In particular, a Sentence Transformer Model and a Sentence Transformer-based Topic Modeling approach were conducted concerning the semantic structure of the textual items [41][42].

The first study by [41] applied the Sentence Transformer Model Augmented SBERT (AugSBERT) [20] to measure the sentence similarity using a cross- and bi-encoder Transformer architecture to encode the measurement items of established UX questionnaires into embedding in a vector space. Afterward, the cosine similarity values between the items were calculated and items were clustered based on a determined threshold. As a result, the similarity clusters containing semantically similar items were identified [41]. The second study extends this approach by applying the specific Topic Modeling technique BERTopic [43] based on the Sentence Transformer SBERT [19]. Therefore, the items were encoded into embeddings in a vector space by applying the SBERT approach. Moreover, the embeddings were clustered using a Topic Modeling technique [42]. Both studies show that innovative NLP techniques can produce plausible results. Nevertheless, there are still several weaknesses in the approaches to be recorded. For further insights, we refer to the respective articles [41][42].

Due to the rapid development of Generative AI, various fields, e.g., NLP are revolutionized [44][45]. Therefore, Generative AI (GenAI) is able to improve processes and contribute valuable results. This study is another approach applying GenAI to find common ground in UX research. We used ChatGPT-4 as LLM [46] to clearly differentiate items semantically and consolidate general factors within established UX questionnaires. The detailed approach is explained in the following section IV.

IV. Methodological Approach

This study applies GenAI for the analysis of UX measurement items. In particular, ChatGPT-4 was used to determine similarity topics based on semantically similar items. The approach is described in the following. ChatGPT-4 is a large multimodal model developed by OpenAI that is able to process data and produce text outputs. The model based on GPT-4 is capable of understanding and generating natural language text [46]. For detailed insights, we refer to OpenAI (https://openai.com/gpt-4).

As a first step in our approach, data was collected. A set of 40 established UX questionnaires [7] was analyzed. We excluded all questionnaires with (1) a semantic differential scale and (2) a divergent measurement concept, i.e., specifically formulated items focusing on a concrete evaluation objective (for further details, see section II-C). This resulted in a list of 19 questionnaires with 408 measurement items. The data collection process is illustrated in Figure 1.

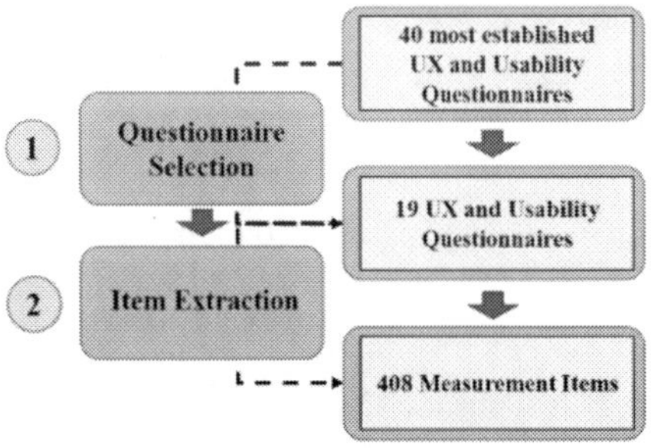

Figure 1: Data Collection.

Secondly, we introduced all items to ChatGPT-4. Thirdly, we formulated seven prompts for ChatGPT-4. The prompts described the task for the LLM. The different tasks given to ChatGPT are described in detail below. The prompts are shown in the following:

- **prompt1**: *"Can you extract the questions with a high similarity, i.e., answering about similar topics?"*
- **prompt2**: *"Can you break this down more detailed?"*
- **prompt3**: *"Can you try to break down each section into more subsections with its own category?"*
- **prompt4**: *"Can you improve your categorization?"*
- **prompt5**: *"In literature, I can find such a list with 16 UX factors.—inserted the defined quality aspects (see Table I)—. Can you compare this list with your categorization and contrast these lists?"*
- **prompt6**: *"I would like you to take your categorization you have done earlier and improve this into more generalized, holistic topics"*
- **prompt7**: *"Below there is a list of statements and questions related to the UX of a software system. Select all statements or questions from this list that describe how easy or difficult it is to learn and understand how to use the software system. List these statements or questions. Start with those statements and questions that describe this best.—inserted list of 408 items from UX questionnaires."*

In relation to ***prompt1***, a simple classification was performed. Based on this, ***prompt2*** should be used for a first extension and development of specific topics. In the next step, the topics were further divided into subcategories ***prompt3***. Further, with ***prompt4*** the task of a topic improvement was specified. For this, the LLM should try to optimize the topics and the respective subcategories classified so far and, thus, create a further advanced classification. Finally, existing UX quality aspects from the literature were introduced to ChatGPT and compared with the AI-generated topics in relation to their similarities and differences ***prompt5***. Until now, we made an exploratory structuring. Moreover, we want ChatGPT to generate and improve the categorizations into more general topics providing a holistic perspective ***prompt6***. Furthermore, we aimed to filter out suitable items that fit a category very well from the existing set of items. We set ***prompt7*** to detect appropriate items using the example of the UX quality aspect ***Learnability***. Such detecting and assignment is particularly useful for "ad-hoc surveys" that do not use a standardized questionnaire to measure UX, but just a bunch of self-made questions to find out something specific. This often requires spontaneous additional questions. Thus, before formulating new items, the search and detection of measurement items within an existing item pool using GenAI is quite practical. Results are shown in the following Section V.

V. RESULTS

In this section, the results of the approach by applying ChatGPT-4 are shown. The sub-sections are aligned to the respective prompts that have been given to ChatGPT.

A. Prompt1: Primary Classification

Referring to the first prompt, the LLM provided a classification by themes and similar topics. This results in six topics. Additionally, the most suitable items have been assigned to each topic. Due to paper restrictions, we have only provided the first three most representative items listed by ChatGPT for each category (see Appendix A1). The classification is shown in the following:

- (1) **Usability and Ease of Use**
- (2) **Design and Aesthetics**
- (3) **User Engagement and Experience**
- (4) **Trust and Reliability**
- (5) **Information Access and Clarity**
- (6) **Issues and Errors**

With regard to the results, common topics emerge. Therefore, functional as well as emotional topics were generated. While observing the items, the topics with their respective items can be considered plausible. Concerning the items, it must be pointed out that the item formulations are very specific, while the different categorizations are very broad in comparison. For example, Topic (1) is named **Usability and Ease of Use**, but the first three representative items refer specifically to Ease of Use. Thus, the respective topics are very broad.

The LLM can identify logical topics based on the semantic textual structure. Nevertheless, as a classification of 6 topics with a total of 408 items seems very superficial, we directly proceeded to the next step. Here we asked the LLM for a more specific classification.

B. Prompt2: More Detailed Classification

We tried to derive a more detailed classification. The respective items are presented in the Appendix (see A2). As a result, ten topics were determined by the LLM.

- (1) **Ease of Use**

- (2) **Complexity and Usability Issues**
- (3) **Design and Appearance**
- (4) **Engagement and Immersion**
- (5) **Performance and Responsiveness**
- (6) **Reliability and Trust**
- (7) **Information Quality and Access**
- (8) **Errors and Bugs**
- (9) **Learning and Memorability**
- (10) **Effectiveness and Efficiency**

Considering the results, the second classification is more precious containing four more topics. Topic (1) in relation to *prompt1* was divided into two topics. Additionally, Performance and Responsiveness, Learning and Memorability, and Effectiveness and Efficiency were introduced. By comparing the results of the first two prompts, the functional, task-related topics were further broken down. Thus, the LLM can distinguish the topics even more precisely. It can be seen that the majority of the AI-generated topics relate to a rather pragmatic quality. Topic (1), (2), (5), (7), (8), (9), and (10) are task-related whereas (3) and (4) address the emotional perception of the user. Topic (6) – Reliability and Trust – contains both task-related and emotional items. Overall, the measurement items seem to be more functionally driven among the topics. Moreover, the item formulation within the different topics is quite broad. Some items can be applied to many scenarios, e.g., *"it meets my needs"*, while others are specified to an application, e.g., *"I feel comfortable purchasing from the website"*. An even more detailed categorization into subcategories therefore seems reasonable.

C. *Prompt3: Extended Classification*

We tried to provide a more detailed classification within each topic and asked for a specific breakdown into subsections. As a result, we obtained 22 further subtopics:

- **Ease of Use**
  System Usability—Website Usability—Application Usability
- **Complexity and Usability Issues**
  System Complexity—Frustration and Difficulty—System Limitations
- **Design and Appearance**
  Visual Attraction—Layout and Structure—Design Consistency
- **Engagement and Immersion**
  Time Perception and Involvement—Depth of Experience
- **Performance and Responsiveness**
  Speed of Response
- **Reliability and Trust**
  Website Trustworthiness—System Reliability
- **Information Quality and Access**
  Quality of Information—Accessibility of Information
- **Errors and Bugs**
  Technical Issues—Error Messages
- **Learning and Memorability**
  Learning Curve—Recall and Retention
- **Effectiveness and Efficiency**
  Functional Efficiency—Expected Functionality

The division into main topics and respective sub-topics confirms that items have the same characteristics on a higher level, but can be further subdivided on a more specific level. This may be due to the different characteristics and focus of the questionnaires and their items. Up to this point, we have determined what categorization levels ChatGPT should take. The next step is to extend ChatGPT to make improvements within its own categorization.

D. *Prompt4: Classification Improvement*

We want ChatGPT to improve the classification without any further specifications. As a result, the LLM identified six main topics with 16 subtopics. For improvement, the number of main topics was reduced which makes it appear that the main topics are again rather broad. This results as well in a broad spectrum of sub-topics. Within the sub-topics, ChatGPT changed the categorizations. For instance, hedonic categories, e.g. **Aesthetics and Design**, are grouped with pragmatic categories, e.g. **Navigation and Usability**. In contrast, the main topic **System Usability and Performance** contains the three sub-topics **Ease of Use, Efficiency and Speed, and Functionality and Flexibility**. Compared to the definition by the DIN ISO [1], the concept of usability is mostly well captured. Concerning the properties, more topics are functional than emotional.

- **System Usability and Performance**
  Ease of Use—Efficiency and Speed—Functionality and Flexibility
- **User Engagement and Experience**
  Engagement Level—Aesthetics and Design—Confusion and Difficulty
- **Information and Content**
  Clarity and Understandability—Relevance and Utility—Consistency and Integration
- **Website-specific Feedback**
  Navigation and Usability—Trust and Security—Aesthetics and Design
- **Learning and Adaptability**
  Learning Curve—Adaptability
- **Overall Satisfaction and Recommendation**
  Satisfaction—Recommendation

In consideration of the results, the categorization improvement emphasizes the two-level structure of the main and sub-topics. However, some main topics are rather broad containing sub-topics with pragmatic as well as hedonic properties.

E. *Prompt5: Comparison Towards Existing Consolidation*

In the following step, we consulted existing UX concepts (see Table I) developed by [6] and compared them to the AI-generated categories. We attempted to draw a comparison between an existing consolidation and the results of the LLM. We defined the prompt as follows: *"In literature, I can find such a list with 16 UX factors.—inserted the defined quality aspects (See Table I) [6]—. Can you compare this list with your categorization and contrast these lists?"*. The comparison is illustrated in Table II:

TABLE II: COMPARISON OF EXISTING UX QUALITY ASPECTS [6] AND AI-GENERATED TOPICS.

(#)	UX Quality Aspects	AI-generated Sub-Topics
(1)	Perspicuity	Ease of Use—Learning Curve
(2)	Efficiency	Efficiency and Speed
(3)	Dependability	Consistency and Integration
(4)	Usefulness	Functionality and Flexibility—Relevance and Utility
(5)	Intuitive use	Ease of Use
(6)	Adaptability	Adaptability
(7)	Novelty	-
(8)	Stimulation	Engagement Level
(9)	Clarity	Clarity and Understandability
(10)	Quality of Content	Relevance and Utility
(11)	Immersion	Engagement Level
(12)	Aesthetics	Aesthetics and Design—Aesthetics and Design
(13)	Identity	-
(14)	Loyalty	Loyalty
(15)	Trust	Trust and Security
(16)	Value	Perceived value

In relation to this comparison, ChatGPT shows some fundamental differences. Firstly, an allocation of the AI-generated topics to all quality aspects is not possible. The factors of *Novelty* and *Identity* stated in the literature [5][6][40] are not covered in the categorization made by ChatGPT. Moreover, there is some overlap between the items and factors as some AI-generated factors can be allocated to more than one quality aspect. Furthermore, the results of the literature (see Table I, [6]) are more generalized. For example, the sub-topic *Trust and Security* is contained in the main topic *Website-specific Feedback*. Hence, *Trust and Security* refers specifically to Websites. In contrast, the quality aspect of *Trust* defined by Schrepp et al. [6] is a main topic of its own described more generally. Thus, existing quality aspects introduce a more holistic view covering both functional and emotional aspects of UX whereas the categorization of the LLM has a stronger focus on the functional side and is more specific. If the categories are too specific, there may be problems with general applicability. Therefore, the objective remains to formulate and present (1) more generally and (2) more emotionally focused categories to provide a universal and holistic perspective towards UX.

F. Prompt6: Construction of Generalized Categories

Against this, we added a further prompt *"I would like you to take your categorization you have done earlier and improve this into more generalized, holistic topics"* to create more generalized topics. In this regard, it is also important to see which items represent the generated topics according to the GenAI as the consolidation and categorization are originally based on the measurement items. We output the top five items representing the respective topic best. As a result, ChatGPT generates a comprehensive overview with generalized UX factors and their definitions. The classification shows a two-dimensional separation into the main topic and sub-topics. Both functional, task-related as well and emotional aspects are contained. This enables a comprehensive and generalized view of the construct of UX made by ChatGPT. The topics and items are shown in the appendix (see A3).

Considering the results, ChatGPT performs very well in consolidating and developing topics concerning a holistic view of UX. Hence, general UX concepts can be derived based on AI-generated topics. Both pragmatic and hedonic dimensions are captured. Mostly, the items are coherent with each other and fit the construct. Especially, functional topics are well generated. However, some weaknesses must be stated. The items differ quite strongly and are accordingly not representative of the respective topic within some categories, e.g. ***Identity***. Moreover, items (4) and (5) categorized in **Consistency and Integration** must be mentioned. The items are clearly of hedonic quality whereas the categorization and other items within the topic are considered pragmatic. Hence, there is a semantic relation between obviously functional and emotional items. For illustration, we have added a **(+)** for a suitable item fit and a **(-)** for an unsuitable item fit in the generated list (see Appendix A3). Additionally, some items may be contained in multiple topics. This can be traced back to the rather general formulation of the measurement items. If this was the case, we added **(+-)**.

G. Prompt7: Searching for Items

Up to this point, we showed how GenAI can be used to exploratively define a semantic structure on a large set of items. Another quite natural use case is to detect those items that represent a clearly defined UX concept. We demonstrate this in the example of the UX concept of learnability (or perspicuity). This concept describes that it is easy to get familiar with a product, i.e. easy to learn and understand how the product can be used [6]. We defined the following prompt *"Below there is a list of statements and questions related to the UX of a software system. Select all statements or questions from this list that describe how easy or difficult it is to learn and understand how to use the software system. List these statements or questions. Start with those statements and questions that describe this best.—inserted list of 408 items from UX questionnaires."*, i.e. an explanation of what we want plus the list of items used as a basis for the analysis.

The resulting list of items contained items that refer to ease of learning *("It was easy to learn to use this system")*, intuitive understanding *("The system was easy to use from the start")*, or aspects that support the user to handle the product *("Whenever I made a mistake using the system, I could recover easily and quickly")*. The top 15 of the resulting items fitted quite well to the request in the prompt (see Appendix A4). Thus, it is relatively simple to use ChatGPT to search for existing items that reflect certain UX concepts. Results indicate a good detection of relevant measurement items concerning the respective UX construct.

VI. Conclusion and Future Work

This article presents a GenAI-based approach for providing a common ground in UX research. We applied the LLM ChatGPT-4 to analyze measurement items concerning semantic similarity from a pool of 408 items related to the most established UX questionnaires. Based on this, ChatGPT-4 generated generalized topics, subtopics, and the respective items. Lastly, ChatGPT detected representative items of existing UX concepts. As a result, six main topics and 15 subtopics were identified. In the following, theoretical and practical implications are drawn.

A. Implications

To conclude, we showed that LLMs can be used to (1) classify items from UX questionnaires concerning their semantic meaning, (2) improve and compare classifications, and (3) detect and assign items to classified topics. Of course, LLMs are inherently non-deterministic models. Thus, if the same sequence of prompts is used again, the resulting classifications will differ. This is in principle not a problem since there is no objectively "correct" classification. If the same task is done independently by several UX experts, the resulting classifications would of course differ too. However, the effort of such an automatic classification is extremely low, and thus the possibility to automatically create several such classifications allows an explorative search for semantic structures in large sets of items that can uncover interesting hidden dependencies that would be hard to detect with a manual analysis by UX experts.

Considering the results, ChatGPT generated a consolidated list of topics, subtopics, and items representing the concept UX comprehensively. Therefore, both functional and emotional aspects were contained. The AI-generated topics indicate a good alignment compared to existing UX concepts. In addition, ChatGPT detected and assigned suitable items to similar topics.

B. Limitations and Future Research

A severe limitation of the paper is that semantic differentials, a quite common item format in UX questionnaires, must be excluded from the analysis to guarantee at least a low level of comparability of the items. Further investigations in prompt engineering must show if it is possible to allow a combination of all common item formats in one analysis.

From a more practical point of view, the results can be used as a measurement framework for quantitative UX evaluation. In future research, a questionnaire for the holistic evaluation of the UX can be compiled from the AI-generated topics and the respective items. Moreover, items from the existing pool could be detected in relation to existing UX concepts and a comprehensive item list for each UX quality aspect can be set up. Such a list for each UX concept can help UX researchers by providing suitable measurement items quickly and easily. Both the questionnaire and the items could be further validated to compromise valid, reliable, and useful results.

This approach is a further step towards a common ground in UX research on the level of the measurement items.

APPENDIX

A1: Respective first three allocated items of AI-generated topics prompt1:

Usability and Ease of Use
The system is easy to use.
I found the system unnecessarily complex.
I thought the system was easy to use.

Design and Aesthetics
The design is uninteresting.
The design appears uninspired.
The color composition is attractive.

User Engagement and Experience
I felt calm using the system.
I was so involved in this experience that I lost track of time.
I lost myself in this experience.

Trust and Reliability
I feel comfortable purchasing from the website.
I feel confident conducting business on the website.
It is a site that feels secure.

Information Access and Clarity
I am able to get the information I need easily.
provides quick and easy access to finding information.
provides relevant information.

Issues and Errors
The system is too inflexible.
The interaction with the system is irritating.
The interaction with the system is frustrating.

A2: Respective first three allocated items of AI-generated topics of prompt2:

Ease of Use
The system is easy to use.
I thought the system was easy to use.
This system is easy to use.

Complexity and Usability Issues
I found the system unnecessarily complex.
I think that I would need the support of a technical person to be able to use this system.
I found this Application X confusing to use.

Design and Appearance
The design is uninteresting.
The design appears uninspired.
The color composition is attractive.

Engagement and Immersion
I felt calm using the system.
I was so involved in this experience that I lost track of time.
I lost myself in this experience.

Performance and Responsiveness
The software responds too slowly.
This software responds too slowly to inputs.
The interaction with the system is fast.

Reliability and Trust
I feel comfortable purchasing from the website.
I feel confident conducting business on the website.
It is a site that feels secure.

Information Quality and Access
I am able to get the information I need easily.
provides quick and easy access to finding information.
provides relevant information.

Errors and Bugs

The system is too inflexible.
The interaction with the system is irritating.
The software has at some time stopped unexpectedly.

Learning and Memorability

I think that I would need the support of a technical person to be able to use this system.
I easily remember how to use it.
It is easy to learn to use it.

Effectiveness and Efficiency

It helps me be more effective.
It meets my needs.
It does everything I would expect it to do.

A3: Generalized AI-generated UX topics with their respective top-5 items:

Usability and Intuitiveness

Ease of Use:
1) The system is easy to use. (+)
2) I think that I would need the support of a technical person to be able to use this system.(+)
3) I would imagine that most people would learn to use this system very quickly.(+)
4) I learned to use it quickly.(+)
5) I can use it without written instructions.(+)

Efficiency and Speed:
1) The interaction with the system is fast.(+)
2) The system responds too slowly.(+)
3) This software responds too slowly to inputs.(+)
4) The speed of this software is fast enough.(+)
5) has fast navigation to pages.(+)

Adaptability:
1) The system is too inflexible.(+)
2) This software seems to disrupt the way I normally like to arrange my work.(+)
3) It is flexible.(+)
4) It requires the fewest steps possible to accomplish what I want to do with it.(+- Efficiency)
5) It is relatively easy to move from one part of a task to another.(+- Efficiency)

Content Quality and Clarity

Relevance and Utility:
1) Provides relevant information.(+)
2) It meets my needs.(+)
3) It is useful.(+)
4) Provides information content that is easy to read.(+)
5) It does everything I would expect it to do.(+)

Consistency and Integration:
1) I thought there was too much inconsistency in this system.(+)
2) I found the various functions in this system were well integrated.(+)
3) I don't notice any inconsistencies as I use it.(+)
4) Everything goes together on this site.(+-)
5) The site appears patchy.(+-)

Clarity and Understandability:
1) The way that system information is presented is clear and understandable.(+)
2) provides information content that is easy to understand.(+)
3) I think the image is difficult to understand.(+)
4) The layout is easy to grasp.(+)
5) I do not find this image useful.(-)

Engagement and Experience

Engagement Level:
1) I was so involved in this experience that I lost track of time.(+)
2) I lost myself in this experience.(+)
3) I was really drawn into this experience.(+)
4) I felt involved in this experience.(+)
5) I was absorbed in this experience.(+)

Stimulation:
1) This experience was fun.(+)
2) I continued to use Application X out of curiosity.(+)
3) Working with this software is mentally stimulating.(+)
4) I felt involved in this experience.(+)
5) During this experience I let myself go.(+- Engagement Level)

Aesthetics and Design:
1) This Application X was aesthetically appealing.(+)
2) The screen layout of Application X was visually pleasing.(+)
3) The design is uninteresting.(+)
4) The layout appears professionally designed.(+)
5) The design appears uninspired.(+)

Trust and Reliability

Trust and Security:
1) I feel comfortable purchasing from the website.(+)
2) I feel confident conducting business on the website.(+)
3) is a site that feels secure.(+)
4) makes it easy to contact the organization.(+)
5) The website is easy to use.(-)

Dependability:
1) This software hasn't always done what I was expecting.(+)
2) The software has helped me overcome any problems I have had in using it.(+)
3) I can recover from mistakes quickly and easily.(+)
4) I can use it successfully every time.(+)
5) Error messages are not adequate.(+)

Novelty and Identity

Novelty:
1) The layout is inventive.(+)
2) The layout appears dynamic.(-)
3) The layout appears too dense.(-)
4) The layout is pleasantly varied.(-)
5) The design of the site lacks a concept.(-)

Identity:
1) Conveys a sense of community.(+)
2) The offer has a clearly recognizable structure.(-)
3) Keeps the user's attention.(-)
4) The layout is not up-to-date.(-)
5) The design of the site lacks a concept.(-)

Value and Loyalty

Perceived Value:
1) I consider my experience a success.(+)
2) My experience was rewarding.(+)
3) The layout appears professionally designed.(+)
4) The color composition is attractive.(+)
5) It is wonderful.(+)

Loyalty:
1) I would recommend Application X to my family and friends.(+)
2) I would recommend this software to my colleagues.(+)
3) I will likely return to the website in the future.(+)
4) I think that I would like to use this system frequently.(+)
5) I would not want to use this image.(+)

A4: Top 15 items filtered for Perspicuity/Learnability
1) It was easy to learn to use this system
2) I could effectively complete the tasks and scenarios using this system
3) I was able to complete the tasks and scenarios quickly using this system
4) I felt comfortable using this system
5) The system gave error messages that clearly told me how to fix problems
6) Whenever I made a mistake using the system, I could recover easily and quickly
7) The information provided with this system (online help, documentation) was clear
8) It was easy to find the information I needed
9) The information provided for the system was easy to understand
10) The information was effective in helping me complete the tasks and scenarios
11) The system was easy to use from the start
12) How the system is used was clear to me straight away
13) I could interact with the system in a way that seemed familiar to me
14) It was always clear to me what I had to do to use the system
15) The process of using the system went smoothly

REFERENCES

[1] I. O. for Standardization 9241-210:2019, *Ergonomics of human-system interaction — Part 210: Human-centred design for interactive systems*. ISO - International Organization for Standardization, 2019.

[2] M. Rauschenberger, M. Schrepp, M. P. Cota, S. Olschner, and J. Thomaschewski, "Efficient measurement of the user experience of interactive products. how to use the user experience questionnaire (ueq).example: Spanish language version," *Int. J. Interact. Multim. Artif. Intell.*, vol. 2, pp. 39–45, 2013.

[3] W. B. Albert and T. T. Tullis, *Measuring the User Experience. Collecting, Analyzing, and Presenting UX Metrics*. Morgan Kaufmann, 2022.

[4] A. Assila, K. M. de Oliveira, and H. Ezzedine, "Standardized usability questionnaires: Features and quality focus," *Computer Science and Information Technology*, vol. 6, pp. 15–31, 2016.

[5] A. Hinderks, D. Winter, M. Schrepp, and J. Thomaschewski, "Applicability of user experience and usability questionnaires," *J. Univers. Comput. Sci.*, vol. 25, pp. 1717–1735, 2019.

[6] M. Schrepp *et al.*, "On the importance of ux quality aspects for different product categories," *International Journal of Interactive Multimedia and Artificial Intelligence*, vol. In Press, pp. 232–246, Jun. 2023. DOI: 10.9781/ijimai.2023.03.001.

[7] M. Schrepp, "A comparison of ux questionnaires - what is their underlying concept of user experience?" In *Mensch und Computer 2020 - Workshopband*, C. Hansen, A. Nürnberger, and B. Preim, Eds., Bonn: Gesellschaft für Informatik e.V., 2020. DOI: 10.18420/muc2020-ws105-236.

[8] H. M. Hassan and G. H. Galal-Edeen, "From usability to user experience," in *2017 International Conference on Intelligent Informatics and Biomedical Sciences (ICIIBMS)*, 2017, pp. 216–222. DOI: 10.1109/ICIIBMS.2017.8279761.

[9] M. Hassenzahl, "The thing and i: Understanding the relationship between user and product," in *Funology: From Usability to Enjoyment*, M. A. Blythe, K. Overbeeke, A. F. Monk, and P. C. Wright, Eds. Dordrecht: Springer Netherlands, 2004, pp. 31–42, ISBN: 978-1-4020-2967-7. DOI: 10.1007/1-4020-2967-5_4.

[10] T. Mikolov, I. Sutskever, K. Chen, G. Corrado, and J. Dean, *Distributed representations of words and phrases and their compositionality*, retrieved: 10/2023, 2013. eprint: 1310.4546. [Online]. Available: https://arxiv.org/abs/1310.4546.

[11] T. Kenter, A. Borisov, and M. de Rijke, *Siamese cbow: Optimizing word embeddings for sentence representations*, 2016. eprint: 1606.04640.

[12] A. Conneau, D. Kiela, H. Schwenk, L. Barrault, and A. Bordes, *Supervised learning of universal sentence representations from natural language inference data*, 2018. eprint: 1705.02364.

[13] Y. Li, D. McLean, Z. A. Bandar, J. D. O'shea, and K. Crockett, "Sentence similarity based on semantic nets and corpus statistics," *IEEE transactions on knowledge and data engineering*, vol. 18, no. 8, pp. 1138–1150, 2006.

[14] H. P. Luhn, "A statistical approach to mechanized encoding and searching of literary information," *IBM Journal of research and development*, vol. 1, no. 4, pp. 309–317, 1957.

[15] K. Spärck Jones, "A statistical interpretation of term specificity and its application in retrieval," *Journal of documentation*, vol. 60, no. 5, pp. 493–502, 2004.

[16] S. E. Robertson, S. Walker, S. Jones, M. M. Hancock-Beaulieu, and M. Gatford, "Okapi at trec-3," *Nist Special Publication Sp*, vol. 109, pp. 109–126, 1995.

[17] S. Deerwester, S. T. Dumais, G. W. Furnas, T. K. Landauer, and R. Harshman, "Indexing by latent semantic analysis," *Journal of the American society for information science*, vol. 41, no. 6, pp. 391–407, 1990.

[18] Q. Le and T. Mikolov, "Distributed representations of sentences and documents," in *International conference on machine learning*, PMLR, 2014, pp. 1188–1196.

[19] N. Reimers and I. Gurevych, "Sentence-bert: Sentence embeddings using siamese bert-networks," in *Conference on Empirical Methods in Natural Language Processing*, 2019.

[20] N. Thakur, N. Reimers, J. Daxenberger, and I. Gurevych, "Augmented sbert: Data augmentation method for improving bi-encoders for pairwise sentence scoring tasks," *arXiv preprint arXiv:2010.08240*, Oct. 2020.

[21] X. Sun *et al.*, "Sentence Similarity Based on Contexts," *Transactions of the Association for Computational Linguistics*, vol. 10, pp. 573–588, 2022, ISSN: 2307-387X. DOI: 10.1162/tacl_a_00477.

[22] I. Gilboa, O. Lieberman, and D. Schmeidler, "Empirical similarity," *The Review of Economics and Statistics*, vol. 88, no. 3, pp. 433–444, 2006.

[23] M. Schrepp, R. Otten, K. Blum, and J. Thomaschewski, "What causes the dependency between perceived aesthetics and perceived usability?," pp. 78–85, 2021.

[24] M. Kuroso and K. Kashimura, "Apparent usability vs. inherent usability, chi'95 conference companion," in *Conference on human factors in computing systems, Denver, Colorado*, 1995, pp. 292–293.

[25] N. Tractinsky, "Aesthetics and apparent usability: Empirically assessing cultural and methodological issues," in *Proceedings of the ACM SIGCHI Conference on Human factors in computing systems*, 1997, pp. 115–122.

[26] W. Ilmberger, M. Schrepp, and T. Held, "Cognitive processes causing the relationship between aesthetics and usability," in *HCI and Usability for Education and Work: 4th Symposium of the Workgroup Human-Computer Interaction and Usability Engineering of the Austrian Computer Society, USAB 2008, Graz, Austria, November 20-21, 2008. Proceedings 4*, Springer, 2008, pp. 43–54.

[27] A. N. Tuch, E. E. Presslaber, M. Stöcklin, K. Opwis, and J. A. Bargas-Avila, "The role of visual complexity and prototypicality regarding first impression of websites: Working towards understanding aesthetic judgments," *International journal of human-computer studies*, vol. 70, no. 11, pp. 794–811, 2012.

[28] C. E. Lance, J. A. LaPointe, and A. M. Stewart, "A test of the context dependency of three causal models of halo rater error.," *Journal of Applied Psychology*, vol. 79, no. 3, pp. 332–340, 1994.

[29] G. T. Ford and R. A. Smith, "Inferential beliefs in consumer evaluations: An assessment of alternative processing strategies," *Journal of consumer research*, vol. 14, no. 3, pp. 363–371, 1987.

[30] D. A. Norman, *Emotional design: Why we love (or hate) everyday things*. Civitas Books, 2004.

[31] D. C. L. Ngo, L. S. Teo, and J. G. Byrne, "Formalising guidelines for the design of screen layouts," *Displays*, vol. 21, no. 1, pp. 3–15, 2000.

[32] G. Bonsiepe, "A method of quantifying order in typographic design," *Visible Language*, vol. 2, no. 3, pp. 203–220, 1968.

[33] I. Díaz-Oreiro, G. López, L. Quesada, and Guerrero, "Standardized questionnaires for user experience evaluation: A systematic literature review," *Proceedings*, vol. 31, pp. 14–26, Nov. 2019. DOI: 10.3390/proceedings2019031014.

[34] B. Laugwitz, T. Held, and M. Schrepp, "Construction and evaluation of a user experience questionnaire," in *HCI and Usability for Education and Work*, A. Holzinger, Ed., Berlin, Heidelberg: Springer Berlin Heidelberg, 2008, pp. 63–76, ISBN: 978-3-540-89350-9.

[35] T. UEQ, "Ueq user experience questionnaire," 2018, retrieved: 10/2023. [Online]. Available: https://www.ueq-online.org/.

[36] H. X. Lin, Y.-Y. Choong, and G. Salvendy, "A proposed index of usability: A method for comparing the relative usability of different software systems," *Behaviour & information technology*, vol. 16, no. 4-5, pp. 267–277, 1997.

[37] M. Schrepp, *User Experience Questionnaires: How to use questionnaires to measure the user experience of your products?* KDP, ISBN-13: 979-8736459766, 2021.

[38] M. Schrepp and J. Thomaschewski, "Design and validation of a framework for the creation of user experience questionnaires," *International Journal of Interactive Multimedia and Artificial Intelligence*, vol. InPress, pp. 88–95, Dec. 2019. DOI: 10.9781/ijimai.2019.06.006.

[39] M. Schrepp, "Ueq+ a modular extension of the user experience questionnaire," 2019, retrieved: 10/2023. [Online]. Available: http://www.ueqplus.ueq-research.org/.

[40] D. Winter, M. Schrepp, and J. Thomaschewski, "Faktoren der user experience: Systematische übersicht über produktrelevante ux-qualitätsaspekte," in *Workshop*, A. Endmann, H. Fischer, and M. Krökel, Eds. Berlin, München, Boston: De Gruyter, 2015, pp. 33–41, ISBN: 9783110443882. DOI: doi:10.1515/9783110443882-005.

[41] S. Graser and S. Böhm, "Quantifying user experience through self-reporting questionnaires: A systematic analysis of sentence similarity between the items of the measurement approaches," in *Lecture Notes in Computer Science, LNCS, volume 14014*, Springer Nature, 2023.

[42] S. Graser and S. Böhm, "Applying augmented sbert and bertopic in ux research: A sentence similarity and topic modeling approach to analyzing items from multiple questionnaires," in *Proceedings of the IWEMB 2023, Seventh International Workshop on Entrepreneurship, Electronic, and Mobile Business*, 2023.

[43] M. Grootendorst, "Bertopic: Neural topic modeling with a class-based tf-idf procedure," *arXiv preprint arXiv:2203.05794*, 2022.

[44] Y. Cao *et al.*, "A comprehensive survey of ai-generated content (aigc): A history of generative ai from gan to chatgpt," *arXiv:2303.04226*, pp. 1–44, 2023, retrieved: 10/2023. [Online]. Available: https://arxiv.org/abs/2303.04226.

[45] A. Bandi, P. V. S. R. Adapa, and Y. E. V. P. K. Kuchi, "The power of generative ai: A review of requirements, models, inputndash;output formats, evaluation metrics, and challenges," *Future Internet*, vol. 15, no. 8, pp. 260–320, 2023, ISSN: 1999-5903. DOI: 10.3390/fi15080260.

[46] OpenAI, "Gpt-4 technical report," *ArXiv*, vol. abs/2303.08774, 2023, retrieved: 10/2023. [Online]. Available: https://arxiv.org/abs/2303.08774.

AI-based Mobile App Prototyping:

Status Quo, Perspectives and Preliminary Insights from Experimental Case Studies

Stephan Böhm and Stefan Graser

CAEBUS Center of Advanced E-Business Studies
RheinMain University of Applied Sciences
Wiesbaden, Germany
e-mail: {stephan.boehm, stefan.graser}@hs-rm.de

***Abstract*—The market for mobile applications is characterized by a large number of applications that are often developed by smaller companies and are distributed free of charge or at low prices via a few central app store platforms. This leads to high innovation rates and high competition. Against this background, a strong customer focus, rapid development, and cost-efficient user-centered design are particularly important for successful mobile apps. For these reasons, prototyping is of great importance in app development. Various mobile app prototyping tools have emerged in recent years, ranging from simple wireframes to high-fidelity prototypes with interfaces for implementing the designed apps. Recent advances in the field of Generative Artificial Intelligence (AI) also offer a wide range of possibilities for assisting and automating app prototyping. Three approaches can be distinguished here: indirect guidance and assistance in prototyping, AI plug-ins as an extension of existing prototyping tools, and innovative prototyping solutions with integrated functionality based on Generative AI. This paper first describes application areas, status quo, and perspectives for using Generative AI in mobile app prototyping. This is followed by describing insights from experimental case studies with selected AI-based mobile app prototyping support. As a result, we demonstrate that simple mock-ups can be generated rapidly with the currently available AI support. While autogeneration of prototypes is more likely to be used for standard use cases, AI support is available for various steps in UX/UI design, which should increase the productivity of app prototyping as a whole soon.**

Keywords–Mobile app prototyping; AI-assisted prototyping; Generative AI; WebAR; ChatGPT.

I. INTRODUCTION

Mobile apps are application software for execution on mobile devices, such as smartphones, with which the functionality of the devices given by hardware and system software can be applied to solve user-specific problems. Typically, mobile apps consist of programs and data installed on the devices by the end users and thus form an important element of device personalization. The introduction of the first mobile app stores around 15 years ago significantly impacted the software market. Since then, a previously not imaginable number of software products have been established and created a new market. Users can select and easily install mobile apps from these markets for almost any purpose. The largest number of mobile apps is available for the Android mobile operating system from Google and the iOS from Apple. As of October 2023, according to [1], nearly 3.8 million such mobile apps were available for users in the Google Play Store and about 1.8 million in the Apple App Store. Many mobile apps are developed by small companies and are offered free of charge, financed by ad revenues, or offered at low prices. This leads to high competition and the need for developers to bring mobile apps into the market quickly, cost-effectively, and closely aligned with user requirements.

For the aforementioned reasons, rapid prototyping, Scrum, or user-centered design (UCD) approaches are very common in mobile app development [2]–[4]. All of these approaches typically start with a phase in which the app idea and basic features are defined by experts and documented as initial requirements. Moreover, in a UCD process, as shown in Figure 1, an attempt is made to involve users in the development process as early as possible to obtain direct feedback about their requirements and preferences [5]. Prototypes are the basis for obtaining this feedback and represent an unfinished state of development of the app concept. They are used to gather user feedback and adapt the prototype to the users' requirements in an iterative process. Since the introduction of mobile apps, more powerful tools for prototyping have been developed. These tools support mobile app designers and developers in transforming their ideas and concepts into prototypes, working on them collaboratively, and presenting them to test users.

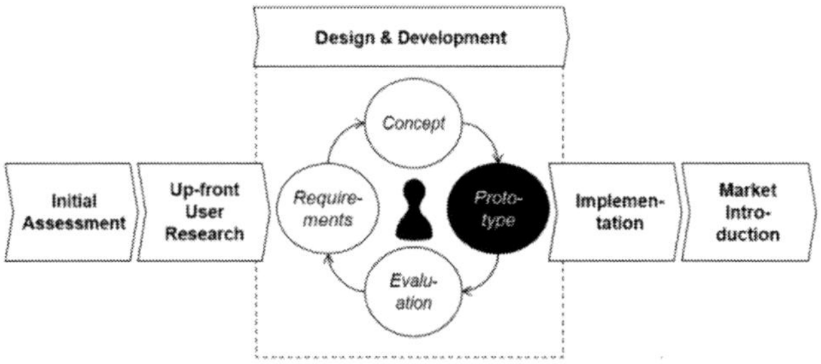

Figure 1. Simplified User Centered Design Process [3].

The fundamental problem of prototyping is to generate demonstrable archetypes from ideas, concepts, and user feedback. This task has typically been performed by screen designers, User Experience (UX) engineers, and app developers. However, prototyping follows experiences and recurring design patterns [6]. Thus, it provides a field of application for supporting and automating activities by new, content-generating forms of artificial intelligence. Such AI-based prototyping support has only recently become available in marketable solutions. This paper will describe the application fields and development status of such AI-based prototyping in an introductory way. Additionally, some first experiences in experimental case studies will be described.

Against this background, the paper is structured as follows: Section 2 gives a brief overview of existing research on the use of AI in mobile app prototyping and formulates the research objectives of this contribution. In the following Section 3, the status quo and perspectives of AI support in mobile app prototyping are discussed. For this purpose, the subject of mobile app prototyping will be specified before the application fields and emerging forms of implementation of AI-based prototyping support are outlined. Section 4 then describes initial experiences and insights from three experimental case studies of AI-based mobile app prototyping support before summarizing the conclusions and implications for practice in Section 5. The paper concludes with a brief outlook on future research topics and needs in Section 6.

II. Related Research and Research Objectives

With the advent of more complex Graphical User Interfaces (GUI) for the web and mobile apps, researchers have been trying to support the laborious prototyping process. As early as 2012, Segura et al. [7] presented the pen-based prototyping tool *UISKEI* for the design of websites, which can recognize certain user interface elements based on rough hand sketches drawn by a designer. An approach to transforming a pixel-based screenshot of a GUI design for mobile apps and web-based technologies into code was presented by Beltramelli [8]. Their *pix2code* approach used machine learning technologies based on convolutional and recurrent neural networks. For graphical user interface designs represented by a single screen, an accuracy of over 77 percent could be demonstrated. Moran et al. [9] present *ReDraw*, a more comprehensive approach for Android combining computer vision, machine learning, and software repository mining to automate prototyping by accurately detecting, classifying, and assembling GUI components. Their approach classified GUI components with a high accuracy of over 90 percent. *ReDraw* generated prototypes close to the mock-ups and a reasonable code structure. Kolthoff et al. [10] proposed *RaWi*, a data-driven GUI prototyping approach. The approach supports Natural Language (NL) searches in a large-scale GUI repository for mobile apps. *RaWi* ranks GUIs from the repository based on advanced machine learning methods (BERT-based LTR models) and provides matches as partly editable GUI screens to support interactive prototyping. Besides research, the potential of AI is also being recognized by providers of prototyping tools. Especially after the introduction of ChatGPT [11] and the increasing popularity of Generative AI, prototyping solutions are now on the market that promise easy prototyping using AI support and highlight the integration of AI or "Powered by AI" in marketing [12]–[14].

In this paper, we can only present selected literature on the state of research. For a detailed description of the state of research, especially on research about GUI and program code retrieval and GUI prototyping in mobile apps, we refer to [10]. However, it can be said that there are already comprehensive approaches in the literature to support the process of visual prototyping for the GUI design of interactive applications, which can be applied to mobile apps or have been developed specifically for this type of software. However, these approaches are mainly based on converting sketches into GUI designs, GUI images into program code, or identifying suitable GUI designs from a repository based on natural language queries. Generating visual prototypes using Generative AI is still an emerging field of research. Against this background, this paper addresses three research questions: (1) In which areas of mobile app prototyping can AI procedures be used, and how can they support the prototyping process? (2) In what form is AI support for mobile app prototyping currently available? (3) What results can be achieved in these areas using AI for exemplary case studies? The paper thus aims to explore the emerging field of AI-based mobile app prototyping and, above all, to generate insights for practice and identify research needs for the future.

III. Mobile App Prototyping and Potential for AI-Assistance

In the following, the process steps and tasks of mobile app prototyping will be described in more detail as a foundation for a more structured discussion of the application potential for AI in mobile app prototyping in the following sections.

A. Mobile App Prototyping

The term prototype comes from the Greek [protos (= the first) and typos (= archetype)] and generally refers to a sample, model, preliminary product, or, more generally, something that is still unfinished [5]. According to Sommerville [15, p. 45], the term prototype in software development refers to "an initial version of a software system that is used to demonstrate concepts, try out design options, and find out more about the problem and its possible solutions."

The basis of mobile app prototypes is usually a significantly shortened requirements elicitation phase compared to conventional software development. In UCD, the initial prototype is defined based on the team's expertise and some limited user research [5]. This phase can use a design-thinking approach and involve a concept formulation phase with semi-structured interviews to gather insights and identify significant elements for the app [16]. These insights are used to formulate the concept for the app's prototype. The prototype is intended to make the mobile app concept and the basic design features and functionalities understandable for test users. In several iterations with feedback loops, the prototype is presented to test users from the target group. The feedback is then evaluated and used to improve the prototype. This process is applied to optimize the design of the prototype in several iterations and to align it as well as possible with the users' requirements.

Figure 2 depicts the different prototyping stages based on a given use case that describes the intended interaction between the user and a system. For a first demonstration of app concepts, visual prototypes are typically used that already represent the intended screen or GUI designs of the concept but otherwise only simulate the system behavior. In contrast to functional prototypes, visual prototypes do not require any coding. However, the actual system behavior can only be tested when at least some system functions have already been implemented. In the case of a working prototype, this implementation is so advanced that an app concept can be tested in field tests in real-world application scenarios.

Typically, the (visual) prototype's abstraction level is adjusted in this iterative process. In the early phases, low-fidelity prototypes are still very different from the later end product.

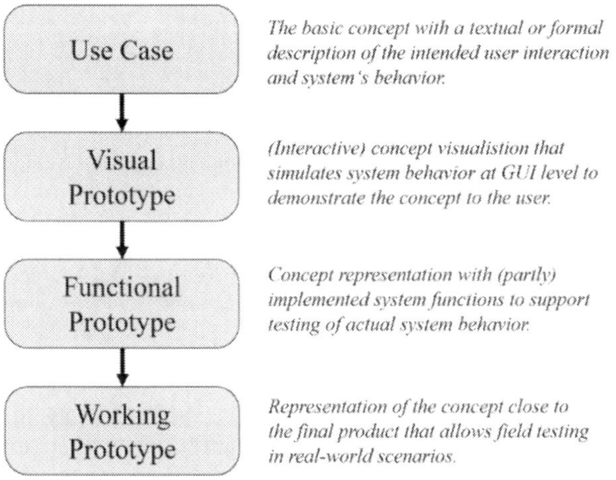

Figure 2. Use Case Definition and Prototypes.

These can be wireframes, for example, initially focusing on the mobile app's basic functionality, screen design, and user flow. Such low-fidelity prototypes show placeholders and wildcards without sophisticated graphical visualization. In later iteration steps, graphical elements are used more intensively, and prototypes are presented to users with a higher fidelity or level of detail. The complexity is further enriched until the app design reflects the intended end product. However, the extent to which app functions are already implemented during this prototyping process can vary depending on the mobile app project and the development approach. Typically, a stable screen prototype is developed first as part of the low and high-fidelity prototyping before the actual implementation and coding of functional components is started. This has the advantage that changes can still be implemented quickly and without unnecessary programming costs.

The prototyping process for mobile applications can be simplified into the following basic steps. Overall, these steps describe prototyping in the broader sense. As mentioned, this understanding of using prototypes to involve users in the development process can be regarded as a UCD approach, as presented in Figure 1. In the narrower sense, prototyping describes creating and refining visual, functional, and working prototypes as presented in Figure 2. The respective phases are briefly described below. For a more detailed discussion, we refer to [5][17].

- *Idea Generation:* Formulation of the app idea and the basic app concept, including the core functionality and features, the intended user value, and the target group. Usage of first design drafts and sketches for screens. At this stage, however, the screen prototypes are usually still isolated, without a screen flow, and do not yet offer interaction options.
- *Basic User Research*: Initial user research and competitive analysis to better understand user requirements, usage context, and own value proposition. For this purpose, focus groups or interviews with potential users can be conducted, or their behavior can be investigated using methods, such as user shadowing or user diaries. However, the app concept is still textual or represented by simple sketches at this stage.
- *Concept Definition:* Description of personas to represent the target audience. Formulation of use cases and the user journey to describe the core interaction between users and the app.
- *Visual Prototyping:* The concept is transferred into interactive visual prototypes. This iterative process starts with simple hand sketches and wireframes in low fidelity. The prototypes are presented to test users to obtain feedback for iterative improvement. Moreover, the level of detail is continuously improved to high-fidelity prototypes.
- *Functional Prototyping:* A functional prototype is generated by programming code for the selected mobile operating systems according to the specifications and the visual prototype. For this purpose, agile software development methods like Scrum break down the app concept into smaller sub-components that can be implemented in a given development time or sprint. The initial functional prototype contains core functions and is progressively enriched with further functionalities.
- *Working Prototype:* The app has already been developed for at least one mobile operating system to the extent that test users can use it in a closed user group outside the app stores to evaluate core functionalities in field tests. Final bugs of the software will be fixed, and the functionality and user experience of the app will be further improved.
- *Continous Improvement:* Launch in the App Store and further improvement through continuous monitoring of user feedback and app reviews in planned updates and long-term versioning. Prototyping can be used to develop and pretest improvements and variants of the app or alternative implementations of sub-functionalities.

The following paper on the support potential of AI is mainly based on prototyping in the narrower sense. It focuses on visual and functional prototypes and how this process can be improved through the use of AI.

B. AI-based Mobile App Prototyping

The discussion of the potential application fields for (Generative) AI for mobile app prototyping is structured based on the phases described above and describes the status quo at the time of writing. The application areas were researched based on a search for case studies, reports, articles, and tool descriptions on AI-based app prototyping. However, as this is an emerging and very new field, there is no relevant, generally accepted term for AI-assisted or -supported prototyping that could be used for research. For a combination of the terms "mobile", "app", "prototyping" with "AI" or "artificial intelligence" in Google Scholar, there was no search result when limited by a search in titles but about 20,700 search results with these terms anywhere in the article in October 2023. Therefore, only an unsystematic explorative search of general Internet sources (e.g., [18]–[22]) could be conducted and systematically summarized. In total, the following seven basic application areas were identified:

1) *Idea Generation and Concept Definition*
2) *Concept Customisation and Refinement*
3) *Design and Mock-up Generation*
4) *Sample Content and Design Variations*
5) *Testing and User Behavior Prediction*

6) *Evaluation of User Feedback and Refinements*
7) *Implementation and Text-to-Code Automation*

These application areas for AI in Prototyping will be described in detail in the following.

1) Idea Generation and Concept Definition: In this early phase of the prototyping process, the customer problem to be solved, the corresponding value proposition, and the central features of the app to be developed are defined and narrowed down. The result can be a product definition statement [5] in which the app concept is initially described in a basic and textual manner. Traditionally, creativity techniques are often used in this phase to develop and further elaborate ideas for app concepts. Generative AI, such as ChatGPT, can be used in this phase to develop ideas for app concepts based on a problem definition and propose central functionalities of such mobile applications. In this process, the Generative AI creates and varies learned concept patterns concerning a given framing. Thus, no completely new app ideas are to be expected. However, creativity technology methods also work with a variation and reconfiguration of known concept elements. By applying Generative AI, a large number of ideas or app variants can be created in a short time.

2) Concept Customisation and Refinement: App concepts are to be adapted to the specific needs of particular target groups. In addition, the app concept must be adapted based on user feedback in an interactive process. Already in early phases and based on textual concept descriptions, potential users can be confronted with the app idea in focus group sessions, for example. Generative AI can simplify the process of editing the concepts accordingly. Instead of manual revision ChatGPT can be asked in a prompt to generate suggestions for adapting the concept. In addition, adaptations to specific target groups can be requested. For this purpose, standardized or common target group definitions can frame the prompts. However, it is also possible to frame the characteristics of a target group independently or to use a previously defined persona for this purpose. Automating customizing for specific target groups or automated refinement based on user feedback can accelerate the maturation of the concept and make this process more efficient. However, experts' expertise or potential users' involvement is still required to evaluate the created variants.

3) Automated Design and Mock-up Generation: The phases described above were still mainly related to textual concepts and thus still represent preparatory phases of the actual mobile app prototyping. As soon as the concept has been sufficiently matured and the essential functionalities and features of the app have been determined, the concept must be transferred into a visual prototype. As described, low-fidelity prototypes are typically created first, e.g., based on wireframes. Generative AI can support transforming textual concepts into visual screen designs. Two steps have to be distinguished. First, the app's overall functionality must be suitably broken down into partial functionalities to derive an efficient user flow with the best possible user experience results. In the second step, the individual screens with their User Interface (UI) elements, such as buttons, input and output fields, labels, and graphic elements, must be designed to be as user-friendly as possible. The combined support of the two steps described above usually requires specialized AI-assisted mobile app prototyping solutions, as discussed below in the form of AI plug-ins for existing prototyping tools or integrated AI-based prototyping solutions (see Section IV).

4) Generation of Sample Content and Design Variations: As discussed earlier, UCD progressively refines the level of detail from low-fidelity prototypes with a still high level of abstraction and iteratively evolves towards high-fidelity prototypes, taking into account user feedback. While low-fidelity prototypes still work predominantly with placeholders, graphical elements, such as icons, images, and texts for labels and descriptions are required later when turning to high-fidelity prototypes. In conventional app design, this is often done by using free or paid content from external sources and providers. In the case of textual content, so-called dummy texts are often used in early phases to fill the text areas on the screens. This text then serves as a pure placeholder and has no meaning related to the app. Generative AI can be used to create prototype content very efficiently in this phase, and it can already be aligned and adapted to the specific app. This not only eliminates the sometimes time-consuming research of corresponding resources but can also save licensing costs for the use of pre-produced third-party content. In addition, variants of corresponding content can be easily generated for user testing. With this form of AI support, it is essential to note that the possibility of directly generating high-fidelity prototypes does not make the use of low-fidelity prototypes obsolete. The high degree of abstraction in low-fidelity prototypes is explicitly used in UCD to initially direct the users' attention and feedback to the essential core functionalities of the app to be developed. Working with a high level of detail too early can distract from the app's core concept and increase the effort –even when using AI– to adapt the prototypes.

5) Automated Testing and User Behaviour Prediction: Once the first screens of a visual mobile app prototype have been created and linked in a screen flow so that user interaction is possible, testing can be conducted to receive user feedback. For this purpose, test scenarios must be worked out, and briefings and tasks for selected users of the potential target group must be created. User feedback is collected and evaluated within the corresponding usability and user experience testing frameworks. For corresponding test procedures, such as A/B testing, it may also be necessary to systematically vary certain design elements of the mobile application to identify the best variant accepted by the users. Such test procedures follow typical processes in which the use of Generative AI can support the preparation phase and the creation of corresponding test scenarios and briefings. Moreover, there are efficiency advantages if test contents and variants can be generated more quickly and automatically, or at least if the AI assists the test managers in generating corresponding materials. In addition, approaches already exist that enable user behavior prediction based on screen designs [23]. The corresponding AI models were trained with the attention distributions or problem areas of example designs. They can then predict corresponding user behavior without testing with real users. However, the precision of such methods and the extent to which such AI support can save user tests even for non-standard designs remains to be further investigated.

6) Automated Evaluation of User Feedback and Refinements: Large amounts of data can accumulate when testing mobile app prototypes. This is the case, for example, when a so-called Thinking Aloud procedure is used, in which users express their experiences and opinions about the prototype in parallel to the testing which is recorded. For evaluation, such recordings are first transcribed from speech into text, for which AI-based methods have been used for some time in the context of Natural Language Processing (NLP). However, with multiple users and more complex test scenarios, large amounts of text can result, that can only be processed manually with a great effort. Here, solutions based on Large Language Models (LLM) can be used to summarise corresponding texts or to extract corresponding problem areas. Procedures based on logging user behavior (e.g., click histories) also generate large amounts of often unstructured or semi-structured data that can be processed and analyzed more efficiently with the help of AI-based procedures.

7) Implementation of Prototypes and Text-to-code Automation: The fields of application so far have been limited to visual prototypes that aim at AI support of the screen design. In such solutions, the actual app functionality is typically still simulated in that the user is forwarded to certain screens depending on a predefined user interaction. Functionalities beyond screen interaction are typically not implemented in visual mobile app prototypes. To implement the functions presented in visual prototypes, programming the app in a Software Development Environment (SDK) of the respective mobile operating system is necessary. In this context, AI-based text-to-code or visual-to-code procedures can be used [13]. This form of AI support can convert (partial) functionalities represented by text prompts or visual screen designs into code and thus into functional or working prototypes. In this phase, there is a transition from prototyping in the narrower sense to software development or technical implementation of the app. However, text-to-code automation is particularly relevant for prototyping when the added value of an app is difficult to simulate or represent through linked interactive "screen dummies". This is the case, for example, when the value for the users depends on interacting with the app in the real world. For example, with mobile Augmented Reality (AR) apps, in which a video stream generated by the device camera is superimposed on the real environment by computer-generated image content. At least part of the AR functionality of the app must already be functionally implemented to give users a realistic impression of the app. In such cases, text-to-code automation using AI could make it possible to create initial functional prototypes without relying on advanced programming skills or corresponding external resources.

IV. Experimental Case Studies on AI-Based Mobile App Prototyping

Most of the previously described support scenarios for mobile app prototyping using AI solutions are only emerging or available in pilots or beta versions in the market. A broader dissemination or research results on the possible efficiency gains with such approaches do not yet exist. Most Internet resources from where the application fields above had been derived had a rather theoretical approach to the AI potentials without referencing a concrete app development project. Against this background, the following section will report on initial experiences with some selected case studies on the application of AI support in mobile app prototyping. Three different forms of AI support are distinguished in the following:

- *AI-based Prototyping Plug-ins.* For some time now, various software solutions have been available to support low- or high-fidelity prototyping of mobile apps. There are solutions specifically for prototyping mobile applications or more universal solutions that, for example, support the creation of prototypes for websites and apps. Additional software or so-called plug-ins from third-party providers can extend some of these solutions. Recently, such AI-based plug-ins have emerged to support app prototyping.
- *Integrated AI-based Prototyping Solutions:* With the advent of Generative AI, the first novel prototyping solutions have become available on the market, highlighting integrated, more comprehensive AI support. Such integrated solutions can, for example, directly transfer textual concepts into visual prototypes. Moreover, further AI-based additional functionalities are embedded in these applications.
- *Prototyping Support by General Generative AI:* Corresponding solutions do not have any specific functionality for mobile app prototyping. However, due to their comprehensive training database and universal applicability, they can be used in different phases of mobile app prototyping. These solutions include the previously mentioned LLMs, such as ChatGPT.

In the following, initial experiences are presented for these three forms of AI support for prototyping based on three selected solutions. For AI-based prototyping plugins, the solution Figma [24] was chosen. Figma is a popular platform for mobile app prototyping characterized by open interfaces and a market with many plug-ins [25]. Appy Pie [12], Uizard [13], and Mockitt [14] were selected as examples of emerging integrated AI-based prototyping solutions emphasizing AI support in their marketing effort. In the area of prototyping support through general Generative AI, ChatGPT [11] was selected because this LLM, launched in the market in November 2022, has the most extensive user base and awareness in the area of Generative AI solutions [26].

Due to the space limitations of this paper, only AI assistance in prototyping for a significantly reduced app concept can be described here. An example from corporate training was defined as a brief use case. The app should explain the functionality of a randomly selected electronic component – a Residual Current-operated Circuit-Breaker (RCCB). This use case was chosen because such a specific solution is uncommon in the app market. Thus, AI solutions cannot simply reproduce existing concepts and design patterns. The following prompt was used for the prototype creation with a very short and simple description that does not list any more concrete functions or content.

Prompt with Use Case Description

Corporate training app for young trainees to learn the function and operation of a residual current circuit breaker via step-by-step illustrated instructions.

In addition to using the plug-ins and integrated solutions for a purely visual app prototype, we tested the application of general Generative AI for text-to-code automation. This AI support was used to create a functional mobile AR prototype, which is not feasible with standard mobile app prototyping tools focusing on GUI design. As mentioned, the following case studies can only allow for initial experiences and preliminary insights into the current state of the market solutions. Based on this, further systematic analyses and research are necessary. However, based on the results, the solutions' status can be presented, and further research needs can be derived.

A. AI-based Prototyping Plug-ins

Numerous prototyping tools are used in mobile app prototyping [27]. Popular app prototyping tools are, for example, Figma, Invision, Mockup, Marvel, and UXPin [28]. These include specialized prototyping tools as well as those that support not only apps but also other interactive applications, such as websites. As already mentioned, the widely used tool Figma, for which thousands of templates, plugins, and UI kits are available [25], will be used as an example of AI support through plugins in this paper. The plugins are available through the Figma community and are provided by independent developers. The plugins can be selected by users of the prototyping platform and integrated into the Figma prototyping environment. At the end of 2023, about 30 plug-ins were available via the community search after entering the terms "artificial intelligence" and "AI". In addition to AI plug-ins for creating, improving, and evaluating prototypes, our search shows tools for task automation in Figma (e.g., naming Figma layers) or prototyping conversational AI applications. We refer to [29] for a more comprehensive description of Figma AI plug-ins for UI/UX design. The support of AI plugins relevant to mobile app prototyping ranges from AI-supported image and text generation to prompt-based creation of wireframes and prediction of user attention distribution on the screens. So far, however, only a few plug-ins support the autodesign or autogeneration of mock-ups based on a text prompt. We identified the Figma plug-in Wireframe Designer [30] that uses the ChatGPT-3.5 API to create mock-ups. Based on a given prompt, ChatGPT creates a design with suitable UI components, converts it into a machine-readable format, and transfers it to Figma for visualization [31]. The three right-hand screens in Figure 3 show the resulting design for the given prompt.

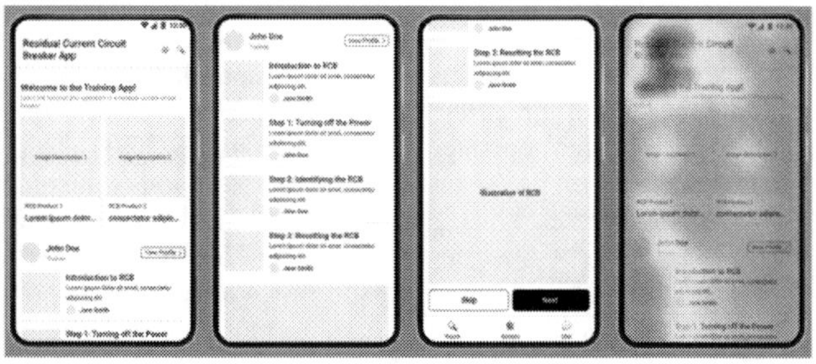

Figure 3. Example Screens and of AI-Generated Prototype with Figma and Plug-ins. [23][24][30].

The heatmap on the right side of Figure 3 was generated by the Attention Insight plugin. The red areas with "warm" colors mark UI elements for which high user attention is predicted. This method substitutes user tests, which usually provide the basis for heatmaps to optimize the screen design. The plugin is based on training data from about 70,000 participants of real heatmap studies. It is supposed to match the results of actual eye-tracking heatmaps for general images with an accuracy of 92.5 percent [32].

Overall, it can be stated that the first working AI plug-ins for existing prototyping solutions are available. Prerequisites are open interfaces with which solutions, such as ChatGPT, can be connected and integrated. In addition, other emerging AI-based plug-ins support individual steps in prototyping up to the assessment of GUI designs.

B. Integrated AI-based Prototyping Solutions

Integrated AI-based solutions for mobile app prototyping are those tools that emphasize AI support in product positioning and offer more integrated AI support for mobile app prototyping. It is to be expected that established popular app prototyping tools will also increasingly integrate AI functions. However, in the second half of 2023, only a few providers were available on the market that positioned themselves with AI-supported prototyping. These include, as already mentioned, Uizard [13], Appy Pie [12], and Mockitt [14]. All of these tools offer an AI-based autodesign, i.e., the creation of an initial mobile app prototype based on the input of a text prompt. Appy Pie can quickly regenerate and process the proposed screen designs in the prototyping environment. Additionally, Uizard offers a wireframe and a screenshot scanner with which hand-drawn sketches and GUI screenshots can be converted into mock-ups. Moreover, design themes, images, and text content can be created and changed, and an attention focus can be predicted. Mockitt offers chat-based interactive support for prototyping. In addition to generating the prototype, flowcharts, mindmaps, and UI components, such as tables, charts, and texts, can be generated with AI support. In addition, further AI functions have been announced on their website [14].

Although the Mockitt application generated a GUI prototype for an app screen from the exemplary prompt, this contained text in Chinese and a random image. Thus, it was not further considered. The result of the Appy Pie tool is shown in Figure 4. The tool generated a straightforward navigation structure with some instructive text content and even added a suitable image. The prototype was simple but sufficient to be refined to gain initial user feedback.

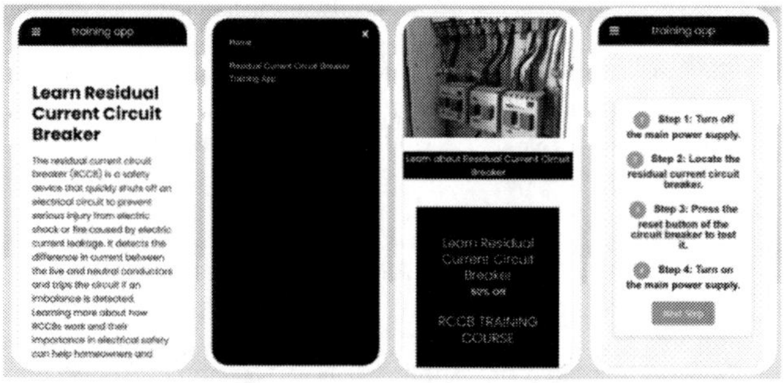

Figure 4. Example Screens of AI-Generated Prototype with Appy Pie [12].

A basic interactive app prototype for the given prompt could also be created with the tool Uizard. Compared to the Appy Pie prototype, more UI elements are added to the GUI design shown in Figure 5. This may be because information on the design style and the target group was also considered. Motivational elements, such as progress bars, and standard components like registration and login masks were integrated into the design. The design was made more colorful and detailed. However, the texts and the images were less specific regarding the defined use case of the RCCB training. Moreover, this prototyping tool did not create steps with instructions for the RCCB based on the prompt.

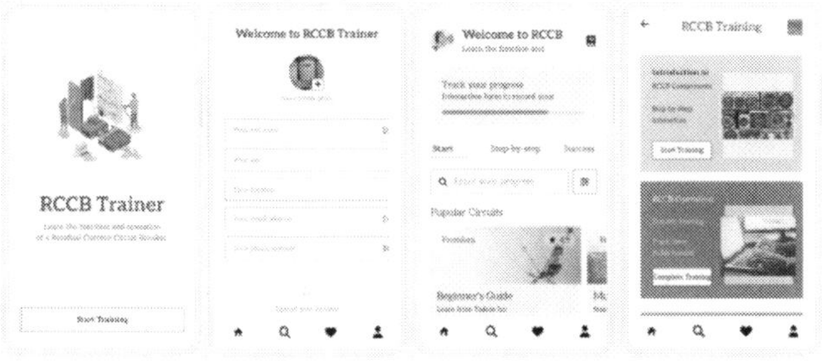

Figure 5. Example Screens of AI-Generated Prototype with Uizard [13].

These examples show that initial designs for mobile app prototypes can already be generated with the current status quo of AI-supported tools but are still in an early (beta) stage. Thus the functions for autogeneration of app designs are more likely to support idea generation for standard app concepts and smaller companies. These approaches cannot yet replace professional app designers. More promising for professional app prototyping are the AI-based productivity tools that allow content to be automatically filled, quickly changed, or even help to switch between different prototyping modes and formats (e.g., hand-drafted wireframes and high-fidelity GUI designs).

C. AI-based Prototyping Support

ChatGPT is a LLM-based chatbot developed by OpenAI and launched in November 2022. Currently, ChatGPT-4 is available for end users [11]. ChatGPT is a chatbot that processes input in natural language through so-called prompts and can be used universally. In addition to outputting text to answer questions, the LLM can generate code in different programming languages and for different software platforms. ChatGPT can generate code for web applications distributed via a web server to be executed in the browser of mobile devices. Due to the simplicity and platform independence of the resulting code and the direct execution without any necessary compilation, corresponding web technologies are well-suited for prototyping. The starting point for this use case was the existence of a 3D model (as a glb-file) of a generic RCCB and the functional specifications below. The research question here was whether a functional web-based AR (WebAR) prototype could be derived from an AI-supported text-to-code approach:

- Automated creation of step-by-step instructions.
- Implementation of the prototype as a WebAR solution (using the AR.js framework [33]).
- Simple button-based navigation between steps.
- Marker-based (Hiro marker) tracking of the RCCB 3D model.
- AR annotations of the 3D object through icon.

To have step-by-step training instructions as a basis for the WebAR app to be generated, the prompt used above was slightly adapted to provide a relevant framing on app prototyping for ChatGPT. A character limit was introduced to avoid lengthy text descriptions.

Adapted Prompt for ChatGPT-4

Prompt: An app designer wants to develop a corporate training app for young trainees to learn the function and operation of a residual current circuit breaker via step-by-step illustrated instructions. Please suggest text descriptions for five instructive steps, each 100 to 250 characters, that can be used as overlays of a WebAR prototype.

ChatGPT successfully processed the creation of step-by-step instructions for a brief training case on RCCBs. The text box below shows the first two of the steps generated. The explanations are very coherently structured and instructive. However, it gets already clear here that the context of an AR application has not been considered further. For example, the second step references a coil, which is only relevant for an AR scenario if internal components are shown. This had to be adjusted manually, as the 3D object available only shows the surface but no inner parts of this component.

ChatGPT-4 Output

1. *Introduction to RCCB:* Welcome to the world of Residual Current Circuit Breakers (RCCB). These devices detect an imbalance between live and neutral currents, ensuring your safety.
2. *Core Components:* RCCBs consist of a few key parts: the coil, the lever, and the trip mechanism. Familiarizing yourself with these will help you understand their operation.
3. ...

In a further step, an attempt was made to generate the code for a prototype WebAR app using ChatGPT. Such a functional prototype allows a realistic impression of the AR view on the smartphone. The technical requirements defined above formed the basis for a prompt. Since a marker-based solution and annotation with icons were requested, the marker to be used and the file names were predefined to avoid the need to subsequently replace corresponding placeholders in the code.

ChatGPT-4 Prompt for Text-to-Code Generation

Provide code to integrate the instructive steps into a WebAR application using the AR.js framework. The instructions for each step should be on a screen with the instructions in a text box on top and navigation buttons at the bottom. in Step 3, the users should be invited to scan a Hiro AR marker to show a 3D model (filename rccd.glb) of the RCCD. In Step 4, the 3D object should be annotated with an icon (filename icon.png).

ChatGPT generated linked HTML, JavaScript, and CSS files as output. The code was initially not executable on the server, mainly due to references to incompatible framework versions, which could be explained by the outdated training data of ChatGPT. However, ChatGPT provided comprehensive support for debugging the code and isolating and eliminating the errors. In addition, many 3D parameters (position, scale, and rotation) had to be adjusted manually to correctly show the 3D object and the annotation on the smartphone screen. Adjusting the parameters and the respective export of the updated code to the server was time-consuming and not an efficient prototyping approach for AR applications. However, in the end, after several iterations of the code, a working WebAR prototype with limited functionality could be derived. Besides the support in debugging the code, the possibilities to improve the screen design by simple text input were helpful. For example, changes to the text size or the display of the text boxes and buttons were requested by prompts, and the code sections to be changed were output by ChatGPT. Figure 6 shows examples of the screens generated with AI assistance by ChatGPT.

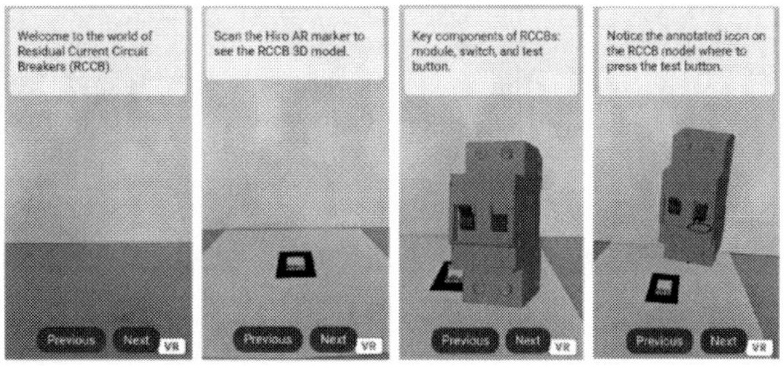

Figure 6. Example Screens of the WebAR Prototype Generated with ChatGPT AI Assistance [11].

In conclusion, it can be said that a simple "functional" AR prototype can be developed using ChatGPT without specific programming knowledge. However, more comprehensive technical knowledge of suitable frameworks is necessary to specify the correct prompts. Also, no directly executable code can be expected, and more comprehensive debugging is necessary, although ChatGPT systematically supports approaches to debugging. The main problem, however, is the very complex positioning of 3D objects and corresponding annotations. ChatGPT was not able to establish relationships between the texts and the objects to be displayed and their properties. Adjusting these parameters iteratively with text prompts is a cumbersome process. Here, graphical AR development environments have clear advantages. On the positive side, however, it should be noted that with ChatGPT's coding assistance, one can familiarise quickly with new frameworks and programming languages and make rapid learning progress.

V. Conclusions and Implications

The applicability of currently available solutions in AI-based prototyping plug-ins, AI-based integrated prototyping solutions, and prototyping support by general Generative AI was investigated using the example of three selected experimental case studies. Almost all solutions used were able to generate corresponding visual or functional prototypes. However, autogeneration of mock-ups produces very basic designs that need further development. The use of general Generative AI requires extensive rework. It is also very complex to apply for AR prototypes due to the lack of a GUI editor or authoring environment if a text-to-code approach is chosen. However, such AI assistance was recognized positively regarding the learning effects imparted. Overall, the state of development is remarkable, considering that the breakthrough of Generative AI was less than a year ago. AI support will likely become widespread in prototyping tools in the next few years. However, the autogeneration of prototypes is probably more relevant for standard applications. More important for productive application in practice will be the AI support of repetitive prototyping sub-tasks, from which a considerable increase in the productivity of UX/UI design processes can be expected.

VI. Outlook on Future Research

This paper is a work-in-progress and presents only preliminary findings on a very new field of research. So far, hardly any papers deal with the possibilities and results of AI support in mobile app prototyping from a deployment perspective. In the past, most papers presented their technical approaches, with which partial aspects of mobile app prototyping can be supported utilizing novel approaches from the AI field. However, marketable AI applications are available, so in addition to the technology, the efficiency of process support and the quality of the results should become a stronger research focus. In the future, it should be investigated to what extent efficiency and productivity advantages can be achieved with such solutions. Of particular importance here is how AI-based prototyping support with general Generative AI such as ChatGPT will evolve, for which a new and more powerful version (ChatGPT-4V(ision) [34]) with enhanced image recognition and code conversion capabilities has just been announced at the time of publication of this article. Furthermore, there is a need for research into the user experience and acceptance of AI-generated solutions compared to conventionally generated and improved prototypes. Given the high number of existing apps in the stores, it is crucial to what extent successful apps with sufficient differentiation potential can be generated with AI or whether the support potential is somewhat limited to standard applications in which AI can increase the efficiency of development processes through a rapid reproduction of proven design patterns. In addition, the AI prototyping tools examined offer support for other formats, such as websites. Unlike apps, these interactive applications are typically characterized by more elaborate graphic designs and complex interaction paths. Thus, further research should investigate how AI prototyping tools provide suitable designs for websites and other interactive applications.

References

[1] 42matters, *Google Play vs the Apple App Store: App stats and trends*, 2023. [Online]. Available: https://42matters.com/stats [retrieved: 10/12/2023].

[2] C. Scharff and R. Verma, "Scrum to support mobile application development projects in a just-in-time learning context," in *Proceedings of the 2010 ICSE Workshop on Cooperative and Human Aspects of Software Engineering*, Y. Dittrich, C. de Souza, M. Korpela, H. Sharp, J. Singer, and H. Winshiers-Theophilus, Eds., New York, NY, USA: ACM, 2010, pp. 25–31, ISBN: 9781605589664. DOI: 10.1145/1833310.1833315.

[3] S. Böhm and B. Igler, "A tool-based approach for structuring feedback for user interface evaluations of mobile applications," *Workshop on Prototyping to Support the Interaction Designing in Mobile Application Development (PID-MAD 2013) in Conjunction with Mobile HCI, Munich, Germany, August 27, 2013,*

2013. [Online]. Available: http://www.hciv.de/pidmad13/proc.html [retrieved: 10/12/2023].

[4] A. K. Almasri, "A proposed hybrid agile framework model for mobile applications development," *International Journal of Software Engineering & Applications*, vol. 7, no. 2, pp. 1–9, 2016, ISSN: 09762221. DOI: 10.5121/ijsea.2016.7201.

[5] S. Ginsburg, *Designing the iPhone User Experience: A User-Centered Approach to Sketching and Prototyping iPhone Apps*. Hoboken: Pearson Education, Limited, 2010, ISBN: 9780321699589.

[6] M. Jurisch, B. Igler, and S. Böhm, "PROFRAME: A prototyping framework for mobile enterprise applications," in *CENTRIC 2016*, L. Berntzen and S. Böhm, Eds., [Wilmington, DE, USA]: IARIA, 2016, pp. 7–10, ISBN: 978-1-61208-502-9.

[7] V. C. V. B. Segura, S. D. J. Barbosa, and F. P. Simões, "UISKEI," in *Proceedings of the International Working Conference on Advanced Visual Interfaces*, G. Tortora, S. Levialdi, and M. Tucci, Eds., New York, NY, USA: ACM, 2012, pp. 18–25, ISBN: 9781450312875. DOI: 10.1145/2254556.2254564.

[8] T. Beltramelli, *Pix2code: Generating code from a graphical user interface screenshot*, May 22, 2017. [Online]. Available: http://arxiv.org/pdf/1705.07962v2.

[9] K. Moran, C. Bernal-Cardenas, M. Curcio, R. Bonett, and D. Poshyvanyk, "Machine learning-based prototyping of graphical user interfaces for mobile apps," *IEEE Transactions on Software Engineering*, vol. 46, no. 2, pp. 196–221, 2020, ISSN: 0098-5589. DOI: 10.1109/TSE.2018.2844788.

[10] K. Kolthoff, C. Bartelt, and S. P. Ponzetto, "Data-driven prototyping via natural-language-based gui retrieval," *Automated Software Engineering*, vol. 30, no. 1, 2023, ISSN: 0928-8910. DOI: 10.1007/s10515-023-00377-x.

[11] OpenAI, *Introducing ChatGPT*, 2023. [Online]. Available: https://openai.com/blog/chatgpt [retrieved: 10/12/2023].

[12] Appy Pie, *AI app generator to generate your app. describe your app in a sentence or two and the AI will help you build it.* 2023. [Online]. Available: https://www.appypie.com/ai-app-generator [retrieved: 10/09/2023].

[13] Uizard, *Design stunning mobile apps in minutes. the world's easiest-to-use design and ideation tool - powered by AI*, 2023. [Online]. Available: https://uizard.io/ [retrieved: 10/09/2023].

[14] Wondershare, *Power your prototyping with Mockitt AI*, 2023. [Online]. Available: https://mockitt.wondershare.com/ai-prototype-generator.html [retrieved: 10/09/2023].

[15] I. Sommerville, *Software engineering*, 9th ed., International ed. Boston: Pearson, 2011, ISBN: 978-0-13-703515-1.

[16] H. M. Isa, R. M. Jusoh, M. H. A. A. Kamal, F. S. M. Amin, and P. F. M. Tamyez, "Enriching user experience among senior citizens in the digital era: A design-thinking approach to constructing a prototype of a mobile application," *Journal of Advanced Research in Business and Management Studies*, vol. 29, no. 1, pp. 20–27, 2022. DOI: 10.37934/arbms.29.1.2027.

[17] B. Bähr, *Prototyping of User Interfaces for Mobile Applications* (T-Labs Series in Telecommunication Services). Cham: Springer International Publishing, Imprint, and Springer, 2017, ISBN: 978-3-319-53209-7.

[18] D. Lane, *How Generative AI will spawn amazing new app ideas*, 2023. [Online]. Available: https://twinsunsolutions.com/blog/how-generative-ai-will-spawn-amazing-new-app-ideas/ [retrieved: 10/12/2023].

[19] Lets Nurture, *Leveraging Generative AI in for seamless mobile app development*, 2023. [Online]. Available: https://www.letsnurture.com/blog/leveraging-generative-ai-in-for-seamless-mobile-app-development.html [retrieved: 10/12/2023].

[20] M. Weaser, *The promise and perils of Generative AI in app development*, 2023. [Online]. Available: https://www.cdotrends.com/story/18381/promise-and-perils-generative-ai-app-development [retrieved: 10/12/2023].

[21] R. Dinakar, *How to use Generative AI to make app testing easy?* 2023. [Online]. Available: https://www.pcloudy.com/blogs/how-to-use-generative-ai-to-make-app-testing-easy/ [retrieved: 10/12/2023].

[22] P. Parra Pennefather, *Creative Prototyping with Generative AI: Augmenting Creative Workflows with Generative AI*. Berkeley, CA: Apress L. P, 2023, ISBN: 9781484295793. [Online]. Available: http://ebookcentral.proquest.com/lib/hsrm/detail.action?docID=30670607.

[23] Attention Insight, *Improve design performance with pre-launch analytics*, 2023. [Online]. Available: https://www.figma.com/community/plugin/968765016617421513/attention-insight [retrieved: 10/10/2023].

[24] Figma, *Work together to build the best products*, 2023. [Online]. Available: https://www.figma.com/design-overview/ [retrieved: 10/10/2023].

[25] Figma, *Welcome to Figma community: Explore thousands of free and paid templates, plugins, and UI kits to kickstart your next big idea.* 2023. [Online]. Available: https://www.figma.com/community/plugins [retrieved: 10/11/2023].

[26] K. Hu, *ChatGPT sets record for fastest-growing user base - analyst note*, 2023. [Online]. Available: https://www.reuters.com/technology/chatgpt-sets-record-fastest-growing-user-base-analyst-note-2023-02-01/ [retrieved: 10/12/2023].

[27] P. Chawla, *List of 28 mobile app design prototyping tools*, 2022. [Online]. Available: https://appinventiv.com/blog/top-mobile-app-prototyping-tools/amp/ [retrieved: 10/11/2023].

[28] H. Clark, *20 best mobile app prototyping tools for product teams in 2023*, 2023. [Online]. Available: https://theproductmanager.com/tools/best-mobile-app-prototyping-tools/ [retrieved: 10/11/2023].

[29] A. Chandak, *10 AI Figma plugins every UI/UX designer must try*, UX Planet, Ed., 2022. [Online]. Available: https://uxplanet.org/10-ai-figma-plugins-every-ui-ux-designer-must-try-d71a9acff1e [retrieved: 10/10/2023].

[30] C. Wu, *Wireframe designer: An AI-powered wireframe generator*, 2023. [Online]. Available: https://www.figma.com/community/plugin/1228969298040149016/wireframe-designer [retrieved: 10/10/2023].

[31] C. Wu, *Over this weekend, I developed a Figma plugin - Wireframe Designer: Twitter post*, 2023. [Online]. Available: https://twitter.com/WCMuu/status/1648164777884266498 [retrieved: 10/10/2023].

[32] Attention Insights, *Attention insights technology*, 2023. [Online]. Available: https://attentioninsight.com/technology/ [retrieved: 10/12/2023].

[33] J. Etienne and N. Carpignoli, *Ar.js - augmented reality on the web*, 2023. [Online]. Available: https://ar-js-org.github.io/AR.js-Docs/ [retrieved: 10/11/2023].

[34] OpenAI, *ChatGPT can now see, hear, and speak*, 2023. [Online]. Available: https://openai.com/blog/chatgpt-can-now-see-hear-and-speak [retrieved: 10/24/2023].

The Multi-Color Contrast Checker (M3C)

Improving visual accessibility of digital interfaces according to WCAG 2.1

Joschua Thomas Simon-Liedtke, Till Halbach
Norwegian Computing Center
Oslo, Norway
e-mail: {joschua,halbach}@nr.no

Abstract—**Accessible and readable contrast of text and graphical elements is a key requirement of the universal design of digital interfaces for people with and without disabilities. Many manual and automatic tools have been developed to help designers and developers measure color contrasts, but these tools are limited when it comes to assessing the contrast of three or more colors, the contrast of non-textual elements, as introduced by the Web Content Accessibility Guidelines (WCAG) 2.1, and contrasts for people with visual deficiencies, including Color Vision Deficiencies (CVDs). This paper proposes an open-source, worldwide accessible, and universally designed web-based tool called Multi-Color Contrast Checker (M3C). Our tool provides three novel main functions: (1) A multi-color contrast checker that assesses two *or more* colors according to their compliance with WCAG 2.1. (2) A color contrast example area that visualizes the effects of the chosen colors on concrete visual example elements typically found in digital interfaces. (3) A CVD simulator for three common types of CVDs. The tool has been assessed for correctness, accessibility, and user experience in both expert and user evaluations, which have confirmed the tool's usefulness in increasing awareness and knowledge of good color contrasts.**

Keywords - Universal design; accessibility; color contrast; Web Content Accessibility Guidelines; WCAG; user evaluation; readability; color vision deficiency; simulation.

I. Introduction

Luminance and color contrast significantly impact the readability and accessibility of visual information in digital interfaces [1][2]. Insufficient contrast affects many user groups, including people with visual impairments, such as people with Color Vision Deficiencies (CVDs) [3][4][5], as well as many elderly [3]. CVDs may prevent users from extracting visual information from a digital interface [5]. Recommendations for accessible contrasts have been integrated into international standards for accessibility, such as the Web Content Accessibility Guidelines (WCAG) [6][7][8][9]. WCAG 2.0 and 2.1 have become part of national and international laws and regulations by requiring compliance with websites, apps, and other ICT solutions [10][11][12]. WCAG defines minimum requirements for the contrast between text and background colors (Guidelines 1.4.3 and 1.4.6 in WCAG 2.0 and later [8][9]) and in non-text elements necessary to understand the content including icons, buttons, parts of a map, charts, etc. (Guideline 1.4.11 in WCAG 2.1 and later [8][9]).

These guidelines aim to make text and graphical elements easy to perceive, ensuring that the content of an application is readable and accessible to individuals with visual impairments. There are numerous application possibilities for WCAG: WCAG can guide designers and developers to choose color combinations with good contrast during development. Likewise, WCAG enables supervisory authorities and interest organizations to assess if colors in an interface are accessible. However, many relevant stakeholders lack awareness and knowledge of WCAG [13][14][15][16]. Moreover, existing color contrast checkers based on WCAG 2.0 [17][18] analyze contrast between two colors (mostly text and background), but lack support for complex multi-colored graphical elements like information graphics, as required by the introduction of non-text content in WCAG 2.1. Existing tools require all relevant color combinations of a multi-colored palette to be checked manually one-by-one. Additionally, there is a lack of example visualization tools that demonstrate how color contrast in complex graphical elements impacts an individual's visual experience of the digital interface. As a result, the success criteria in WCAG can remain somewhat theoretical, leading to developers and designers not adequately considering individuals with visual variations during the design and development process. Last, the needs of individuals with CVDs are often overlooked, even though research shows that CVD simulations can enhance understanding of the experience of people with CVDs among individuals with normal color vision [19][20].

We aim to enhance the accessibility of visual information in digital interfaces by facilitating compliance with the color contrast success criteria defined in WCAG 2.1. We present a web-based tool to support developers, designers, decision-makers, authorities, and civil society organizations in analyzing and visualizing accessible contrasts for simple and complex color combinations that meet the WCAG 2.1 requirements. Additionally, the tool increases awareness of the most common types of CVD. Compared to other tools our solution supports combinations of more than two colors, provides typical UI examples, and simulates the most common CVDs for the chosen colors.

In this paper, we discuss in Section II related color checkers. Section III presents the concept and implementation of the Multi-Color Contrast Checker (M3C). Section IV introduces methods to assess the universal design of the developed tool. In Section V, we discuss their results including some suggestions for future research and development before concluding the article in Section VI.

II. Related solutions

The W3C maintains a list of tools that can measure color contrasts in digital interfaces [21] that possess some notable limitations. Only a few tools compute contrast between multiple colors like, for example, Accessible Brand Colors [22]. Many color checkers only calculate contrast between two colors like, for example, Color Check [23], Clapperton's Colour Contrast Checker [24], Monsido's Color Contrast Checker [25], Color Contrast Checker by UserWay [26], Colors Tester [27], ColorTester [28], Rumoroso's WCAG Contrast Checker [29], WCAG Contrast Checker by Acart Communications [30], contrast finder [31], Contrasts - WCAG Farbkontraste [32], Color Contrast Accessibility Validator [33], Visual Contrast Checker and Colorblind Simulator [34]. In addition, most of these contrast checkers support only textual elements and their backgrounds [22][23][24][27][28][30][31]. A few color checkers provide CVD simulation [24][32], while others can *only* be used for CVD simulation [35][36]. Some tools do not have an explicit connection to the success criteria of WCAG [23][30][31], while some have only limited free functions [27][34] or are limited to specific operative systems or apps [29][32].

Popular tools not on the W3C list include WebAIM's Contrast Checker [18], Colour Contrast Analyser [37], Color Contrast Checker & Analyzer [38], and Firefox' native Accessibility Inspector [39]. Three of them check for both text and non-textual elements [18][37][39], and only one includes a simulation of different CVD types [39].

Many existing tools calibrated for calculating contrast, specifically between text and background colors per WCAG 2.0 guidelines, fall short of accommodating the expanded color usage for non-text elements in user interfaces introduced by WCAG 2.1. The current tools largely neglect practical visualization of color combinations in various graphical interface elements and lack consideration for color-deficient users. Additionally, tools offering broader functionalities are generally neither universally available nor commercially free. Hence, there is a notable absence of an open-source tool that effectively enables computation and practical demonstration of multiple color contrasts, while also providing CVD simulations. Moreover, there is a necessity for scientific research on strategies to ensure that such a tool is universally designed.

III. The Multi-Color Contrast Checker (M3C)

To facilitate and increase compliance with WCAG 2.1, we propose a web-based tool, called *Multi-Color Contrast Checker (M3C)*, which is openly available at https://norskregnesentral.github.io/m3c/, easy to understand and use, accessible, and universally designed for people with and without disabilities. The tool consists of a multi-color contrast checker, a color contrast example area, and a CVD simulator.

A. The Multi-Color Contrast Checker

The multi-color contrast checker (cf. Figure 1) calculates contrasts between multiple colors. The user can make a color palette by choosing colors found in a text, background, non-textual element, or graphics manually from a color wheel (in RGB or HSV), by entering a hex color value, or by using a pipette on a graphical element visible on the screen. The contrast checker enlists all possible combinations of the chosen colors and assesses whether each combination meets the success criteria 1.4.3, 1.4.6, and 1.4.11, as defined in WCAG 2.1 for texts (both minimum and enhanced), as well as non-textual elements [8]. Information about WCAG 2.1 and color contrast according to WCAG 2.1 is included in the tool. The numerical results of the contrast computations are presented in a matrix, along with indicators for whether the success criteria are met on level AA or AAA (cf. Figure 1). Contrast computations are defined in WCAG 2.1 [8]:

$$(L_1 + 0.05) / (L_2 + 0.05) \quad (1)$$

Where L_1 and L_2 represent the relative luminance of the lighter and darker colors respectively according to the sRGB standard [40][41]. The tool also includes a "Get started" section that offers accessible example color palettes inspired by Color Brewer 2.0 and Color Hunt [42][43][44].

B. The Color Contrast Example Area

The color contrast example area (cf. Figure 2) demonstrates the practical implications of contrast and readability of the chosen colors. It provides concrete examples of typical UI elements and visualizes how distinct color combinations affect these elements' readability. The example area uses vector graphics that can individually adapt the colors of the various parts of the graphic. In the current implementation, we included four UI elements: text on background, a calendar graphic, a pie chart, and a button.

C. The Color Vision Deficiency Simulator

The Color Vision Deficiency (CVD) simulator (cf. Figure 1) replicates the three most common types of CVDs, protanopia, deuteranopia, and tritanopia, based on the formulas provided by Brettel et al. [45][46]. The CVD simulator automatically adjusts the colors of the palette and in the example area to mirror perceptions of a person with CVD. Concurrently, all contrast calculations are updated to reflect the simulated colors, enhancing the understanding of contrast perception for a person with CVD.

D. Implementation

The code of the application is hosted on and deployed by GitHub [47][48]. The tool has been programmed in the JavaScript library React [49] including the libraries i18next for internationalization [50], color-contrast-checker for contrast calculations [51], and adaptations of libDaltonLens for CVD simulations [52][53]. The CSS libraries flexbox and grid enabled a responsive user interface for variable screen sizes. The tool has been evaluated on computer screens, mobile phones, and tablets [54]. The tool has been developed in an agile process with several iterations consisting of planning, execution, review, and retrospective [55] including a review by other developers during each cycle. We conducted a round of user evaluation aimed at accessibility and user experience on the first prototype in January 2023.

IV. Accessibility and User Experience Evaluations with Experts and Users

A. Methods and Participants

We used tools, checklists, and user evaluations to assess the accessibility and user experience of the web tool. To begin with, the code was periodically tested concerning rudimentary technical accessibility using the Chrome and Firefox extensions of the WAVE Web Accessibility Evaluation tools [56]. Next, we conducted an expert evaluation according to the WCAG 2.2, including screen reader testing. We chose evaluation using WCAG 2.2 to make the tool more robust for the future, due to WCAG's backward compatibility with previous versions.

We conducted user evaluations of our first prototype in January 2023. This prototype contained all functions on a single page, requiring the user to scroll up and down to each section. The prototype also lacked the CVD simulation and internationalization modules. The evaluations were inspired by inclusive-design approaches and strategies for the assessment of digital artifacts' user experience [57][58][59][60]. The evaluations employed a think-aloud strategy to identify bugs and other issues [57][60], and we defined a set of tasks that the participants had to complete in the presence of the test leader. Practical, technical, and content-related tasks were included, aiming at the functionality, accessibility, usability, and user-friendliness of the tool. The test leader encouraged the participants to comment while completing the tasks. Where necessary, the test leader asked clarifying questions. Additionally, we interviewed the participants about their expectations, their first impressions, and impressions after testing, including grades for satisfaction, the tool's easy-to-use factor, and perceived usefulness.

User evaluations were conducted with one female and seven male developers, both locally (5) and remotely (3) using our experience from previous remote evaluations [61]. Seven of the participants did not mention any disabilities, while one reported some degree of low vision. Some of the participants had corrected-to-normal vision otherwise. The average level of experience with universal design among the participants was low (2.4 on a 5-point Likert scale). We evaluated multiple operating systems including Windows (6), Linux (1), and macOS (1), as well as different browsers including Chrome (4), Firefox (3), and Edge (1).

B. Results

The WCAG testing resulted in the discovery of 12 accessibility issues. The expert and automated testing resulted in the detection of 61 issues in total, of which 41 were usability issues, 13 were accessibility issues, and 7 were other issues. The issues were then triaged together with the accessibility issues and mostly implemented.

The participants reported a medium satisfaction grade (3.25 on a 5-point Likert scale). They reported that the first version of the prototype was neither / nor easy-to-use (3 on a 5-Point Likert scale). Finally, participants gave the tool a high usefulness rating (4.25 on a 5-point Likert scale).

Participants found the prototype of the tool lacking in essential information about its purpose, funding, and creators, and criticized the absence of guidance and tutorials making it non-intuitive and difficult to comprehend. The front page was deemed cluttered and overwhelming with numerous elements, and the interface's bright, sharp colors made it appear unrefined.

Participants praised the interface for its clean, spacious design, clear hierarchy, and color categorization of the sections, appreciating the immediate, universal color changes and well-categorized sections. The color contrast calculations matrix and interactivity of the example area were valued for providing clear overviews and hands-on visual manipulations, respectively. The summary feature highlighting good and bad color contrasts and the presentation of WCAG requirements were deemed useful, with both the matrix and summary commended for showing all possible color combinations.

Participants recommended making the system more logical by utilizing unused space, emphasizing essential elements, and refining icon usage. Suggestions included implementing tabs to minimize scrolling and optionally presenting WCAG guidelines alongside the contrast ratio table. The need for multi-language support and a feedback mechanism for reporting issues and suggesting improvements was emphasized. Additionally, participants desired a "dark mode" and increased interactivity in the example area with more than two colors, as well as guidance for choosing accessible color combinations from the start.

V. Discussion

The tool's accessibility and user experience have considerably improved for the current version. Tabs have been added to make the app's structure clearer by avoiding empty space, giving clear headers, and putting focus on a single section at a time while hiding others. We updated the WCAG explanation and placed it in a separate tab for clarity. We integrated internationalization and translations in English. We added background information about the project and its contributors and added the possibility to submit feedback through GitHub's issue feature. Last, we implemented the CVD simulation feature.

User feedback generally indicated a positive reception, with users gaining familiarity with WCAG and appreciating the insights offered by the new CVD feature in understanding the needs of individuals with CVDs.

The proposed tool targets developers, designers, decision-makers, regulatory authorities, and civil society organizations. The tool can support these stakeholders' work by assessing if a chosen palette for a digital user interface conforms with the WCAG 2.1 success criteria. In practice, the tool can be utilized for all digital interfaces with visual information such as websites, web applications, digital learning materials, and self-service machines, as well as computer games, XR apps, or IoT devices, but this list is non-exhaustive. The tool also promotes knowledge and awareness of the universal design of ICT applications by providing information about good color contrasts and WCAG. The application is built modularly, allowing for easy expansion, and adding new features, such as support for additional languages, simulations, or more graphical examples.

The tool underwent user experience and accessibility evaluations including WCAG checks and user testing to ensure that it is universally designed. It can, for example, be operated by only using the keyboard or using a screen reader. We also utilized the tool to assess the accessibility of the colors employed on the website itself. The application is deployed and freely available on GitHub, enabling all users worldwide to benefit from the tool. Also, the source code may be downloaded and extended as desired. Users can provide feedback and report issues or requests for future features. At the same time, the current research has its limitations related to the lack of quantitative performance comparisons with other tools, as well as demonstrating real-world impact by including user studies or metrics quantifying improvements in accessibility and WCAG compliance.

We aim to enhance our website's accessibility and user guidance by implementing several upgrades: translating it into additional languages, including German, Chinese, Nynorsk, and Sámi, to appeal to a global audience; refreshing and augmenting examples; and introducing a tutorial through a succinct video demonstration. Additionally, we plan to incorporate a color palette picker, designed to facilitate accessible color selection in compliance with WCAG 2.1 while adhering to other functional and aesthetic needs—ensuring it meets user expectations related to, for instance, corporate design, and allowing user control over parameters such as hue, saturation, and brightness. Furthermore, we will assess the tool's efficacy and its contribution to increasing WCAG knowledge and awareness through questionnaires and head-to-head comparative evaluations, thereby establishing its innovations and added value compared to existing tools.

VI. Conclusion

We introduced the Multi-Color Contrast Checker (M3C), a novel tool designed to enhance the accessibility and readability of visual information in adherence to WCAG 2.1 contrast success criteria. M3C, which is openly available and free-to-use (1) enables multi-color contrast checking, (2) visualizes color contrast effects on graphical interface elements, and (3) simulates various types of Color Vision Deficiency (CVD). Besides aiding developers and designers in making informed color selections, it also facilitates assessments of color palettes against WCAG 2.1 criteria.

We employed diverse evaluation methods, including user experience and accessibility testing, to ensure that the tool itself is universally designed. Future developments will incorporate additional languages and a color palette chooser, guiding users toward accessible color selections within existing aesthetic, functional, or corporate design constraints.

Acknowledgment

This work has been supported by the UnIKT program of the Norwegian Directorate for Children, Youth, and Family Affairs (Bufdir). We thank Kamilla Mortensen and Thor Kristoffersen who helped with the development of the tool.

References

[1] K. Knoblauch, A. Arditi, and J. Szlyk, "Effects of chromatic and luminance contrast on reading," *J. Opt. Soc. Am. A*, vol. 8, no. 2, pp. 428–439, Feb. 1991.

[2] B. A. Parker and L. F. Scharff, "Influences of contrast sensitivity on text readability in the context of graphical user interface," *Unpublished manuscript. Retrieved December*, vol. 20, p. 2007, 1998.

[3] G. Haegerstrom-Portnoy, M. E. Schneck, and J. A. Brabyn, "Seeing into old age: vision function beyond acuity," *Optom. Vis. Sci.*, vol. 76, no. 3, pp. 141–158, Mar. 1999.

[4] C. Rigden, "'The Eye of the Beholder'-Designing for Colour-Blind Users," *British Telecommunications Engineering*, vol. 17, pp. 291–295, 1999.

[5] B. L. Cole, "The handicap of abnormal colour vision," *Clin. Exp. Optom.*, vol. 87, no. 4–5, pp. 258–275, Jul. 2004.

[6] A. Arditi and K. Knoblauch, "Choosing effective display colors for the partially sighted," *SID International Symposium Digest of Technical Papers*, vol. 25, pp. 32–32, 1994.

[7] A. Arditi and K. Knoblauch, "Effective color contrast and low vision," *Functional Assessment of Low Vision*, pp. 129–135, 1996.

[8] World Wide Web Consortium (W3C), "Web Content Accessibility Guidelines (WCAG) 2.1," Jun. 2018. https://www.w3.org/TR/WCAG21/ (retr. Oct., 2023).

[9] World Wide Web Consortium (W3C), "Web Content Accessibility Guidelines (WCAG) 2.2," May 2023. https://www.w3.org/TR/WCAG22 (retr. Oct., 2023).

[10] Kommunal- og moderniseringsdepartementet, "Regulation on Universal Design of Information and Communication Technology (ICT) Solutions (FOR-2013-06-21-732)," original: "Forskrift om universell utforming av informasjons- og kommunikasjonsteknologiske (IKT)-løsninger (FOR-2013-06-21-732)," Jun. 2013. https://lovdata.no/dokument/SF/forskrift/2013-06-21-732 (retr. Oct., 2023).

[11] European Union (EU), "Web Accessibility Directive (WAD) - Directive (EU) 2016/2102," Oct. 2016. https://eur-lex.europa.eu/eli/dir/2016/2102/oj (retr. Oct., 2023).

[12] European Telecommunications Standards Institute (ETSI), "EN 301 549 v3.2.1 (2021-03): Accessibility requirements for ICT products and services." Mar. 2021. [Online]. Available: https://www.etsi.org/deliver/etsi_en/301500_301599/301549/03.02.01_60/en_301549v030201p.pdf (retr. Oct., 2023).

[13] World Wide Web Consortium (W3C), "Understanding success criterion 1.4.3: Contrast (minimum)," 2016. https://www.w3.org/WAI/WCAG21/Understanding/contrast-minimum.html (retr. Oct., 2023).

[14] Y. Inal, F. Guribye, D. Rajanen, M. Rajanen, and M. Rost, "Perspectives and Practices of Digital Accessibility: A Survey of User Experience Professionals in Nordic Countries," in *Proc. of the 11th Nordic Conference on Human-Computer Interaction: Shaping Experiences, Shaping Society*, in NordiCHI '20, no. 63. New York, NY, USA: Association for Computing Machinery, Oct. 2020, pp. 1–11.

[15] J. T. Simon-Liedtke and R. C. Baraas, "The Future of eXtended Reality in Primary and Secondary Education," *Stud. Health Technol. Inform.*, vol. 297, pp. 549–556, Sep. 2022.

[16] Tilsynet for universell utforming av IKT (uutilsynet), "Suggestions for websites – contrast," original: "Løsningsforslag for nettsider – Kontrast," 2023. https://www.uutilsynet.no/wcag-standarden/kontrast/48 (retr. Oct., 2023).

[17] Accessible Web, "WCAG Color Contrast Checker," Mar. 16, 2021. https://accessibleweb.com/color-contrast-checker/ (retr. Oct., 2023).

[18] Utah State University - Institute for Disability Research, Policy, and Practice, "WebAIM: Contrast Checker," 2023. https://webaim.org/resources/contrastchecker/ (retr. Oct., 2023).

[19] D. R. Flatla and C. Gutwin, "'So that's what you see': building understanding with personalized simulations of colour vision deficiency," in *Proc. of the 14th int. ACM SIGACCESS conference on Computers and accessibility*, in ASSETS '12. New York, NY, USA: Association for Computing Machinery, Oct. 2012, pp. 167–174.

[20] G. W. Tigwell, "Nuanced Perspectives Toward Disability Simulations

from Digital Designers, Blind, Low Vision, and Color Blind People," in *Proc. of the 2021 CHI Conference on Human Factors in Computing Systems*, in CHI '21, no. 378. New York, NY, USA: Association for Computing Machinery, May 2021, pp. 1–15.

[21] World Wide Web Consortium (W3C), "Web Accessibility Evaluation Tools List," 2022. https://www.w3.org/WAI/ER/tools/ (retr. Oct., 2023).

[22] Use All Five, "Accessible Brand Colors," 2023. https://abc.useallfive.com/ (retr. Oct., 2023).

[23] Off-Site Services, Inc., "Color Check for ADA image compliance," 2023. https://www.oss-usa.com/color-check-ada-image-compliance (retr. Oct., 2023).

[24] A. Clapperton, "Color Contrast & Color Perception Accessibility Checker," 2018. https://www.websiterating.com/color-contrast-perception-checker/ (retr. Oct., 2023).

[25] Monsido, "Color Contrast Checker," 2023. https://monsido.com/tools/contrast-checker (retr. Oct., 2023).

[26] UserWay, "Color Contrast Checker," 2023. https://userway.org/contrast/ (retr. Oct., 2023).

[27] MediaAndMe, "Colors Tester," 2023. https://colors-tester.com/ (retr. Oct., 2023).

[28] Alfasado, "ColorTester," 2020. https://alfasado.net/apps/colortester.html (retr. Oct., 2023).

[29] Rumoroso, "WCAG Contrast checker," May 25, 2008. https://addons.mozilla.org/en-US/firefox/addon/wcag-contrast-checker/ (retr. Oct., 2023).

[30] Acart Communications, Inc., "WCAG Contrast Checker," Sep. 27, 2017. https://contrastchecker.com/ (retr. Oct., 2023).

[31] Tanaguru, "Tanaguru Contrast-Finder," 2023. https://contrast-finder.tanaguru.com/ (retr. Oct., 2023).

[32] C. Wendt, "Contrasts - WCAG Color Contrasts," original: "Contrasts - WCAG Farbkontraste," 2023. https://apps.apple.com/de/app/contrasts/id1515015989 (retr. Oct., 2023).

[33] Bureau of Internet Accessibility (BoIA), "Color Contrast Accessibility Validator." https://color.a11y.com/?wc3 (retr. Oct., 2023).

[34] Adee, "Web accessibility testing tools and platform," 2023. https://adee.co/ (retr. Oct., 2023).

[35] B. Jenny, "Color Oracle," 2006. https://colororacle.org/ (retr. Oct., 2023).

[36] Ryobi Systems Co., Ltd., "Visolve," 2020. https://www.ryobi.co.jp/products/visolve/en/ (retr. Oct., 2023).

[37] TPGi, "Color Contrast Analyzer," Jan. 24, 2021. https://www.tpgi.com/color-contrast-checker/ (retr. Oct., 2023).

[38] SBF Consulting, "Color Contrast Checker / Analyzer," 2013. https://www.sbwfc.co.kr/color-contrast-checker/ (retr. Oct., 2023).

[39] Mozilla Foundation, "Firefox Accessibility Inspector," 2023. https://firefox-source-docs.mozilla.org/devtools-user/accessibility_inspector/index.html (retr. Oct., 2023).

[40] M. Stokes, M. Anderson, S. Chandrasekar, and R. Motta, "A Standard Default Color Space for the Internet - sRGB," Nov. 05, 1996. https://www.w3.org/Graphics/Color/sRGB.html (retr. Jul. 11, 2023).

[41] Int. Electrotechnical Commission (IEC), "Multimedia systems and equipment - Colour measurement and management - Part 2-1: Colour management - Default RGB colour space - sRGB," IEC 61966-2-1:1999, Oct. 1999. retr.: Jul. 11, 2023. [Online]. Available: https://webstore.iec.ch/publication/6169

[42] M. Harrower and C. A. Brewer, "ColorBrewer.org: An Online Tool for Selecting Colour Schemes for Maps," *Cartogr. J.*, vol. 40, no. 1, pp. 27–37, Jun. 2003.

[43] C. Brewer, M. Harrower, B. Sheesley, A. Woodruff, and D. Heyman, "ColorBrewer 2.0," *ColorBrewer: Color Advice for Maps*, 2002. https://colorbrewer2.org/ (retr. Jul. 11, 2023).

[44] G. Shir, "Color Hunt," *Color Hunt - Color Palettes for Designers and Artists*, 2015. https://colorhunt.co/ (retr. Jul. 14, 2023).

[45] H. Brettel, F. Viénot, and J. D. Mollon, "Computerized simulation of color appearance for dichromats," *J. Opt. Soc. Am. A Opt. Image Sci. Vis.*, vol. 14, no. 10, pp. 2647–2655, Oct. 1997.

[46] J. T. Simon-Liedtke, "Assessment and Design of Color Vision Deficiency Simulation and Daltonization Methods," Ph.D., Norges teknisk-naturvitenskapelige universitet (NTNU), 2017. [Online]. Available: https://ntnuopen.ntnu.no/ntnu-xmlui/handle/11250/2440785 (retr. Oct., 2023).

[47] J. T. Simon-Liedtke, K. Mortensen, T. Halbach, and T. Kristoffersen, "The Multi-Color Contrast Checker (M3C) - deployment," Aug. 05, 2022. https://norskregnesentral.github.io/m3c/ (retr. Oct, 2023).

[48] J. T. Simon-Liedtke, K. Mortensen, T. Halbach, and T. Kristoffersen, "The Multi-Color Contrast Checker (M3C) - code". Github, 2022. retr. [Web-based]. Available: https://github.com/NorskRegnesentral/m3c (retr. Oct, 2023).

[49] Meta Open Source, *react*. Github, 2022. [Online]. Available: https://github.com/facebook/react/releases?page=10 (retr. Oct., 2023).

[50] A. Raiano and J. Mühlemann, *i18next*. Github, 2023. [Online]. Available: https://github.com/i18next/i18next (retr. Oct., 2023).

[51] C. Turansky, *color-contrast-checker*. Github, 2020. [Online]. Available: https://github.com/nikaocreatives/color-contrast-checker (retr. Oct., 2023).

[52] N. Burrus, "Understanding LMS-based Color Blindness Simulations," Oct. 21, 2021. https://daltonlens.org/understanding-cvd-simulation/ (retr. Oct., 2023).

[53] N. Burrus, *libDaltonLens*. Github, 2021. Available: https://github.com/DaltonLens/libDaltonLens (retr. Oct., 2023).

[54] World Wide Web Consortium (W3C), "CSS Snapshot 2021," Dec. 31, 2021. https://www.w3.org/TR/css-2021/ (retr. Oct., 2023).

[55] M. Drury, K. Conboy, and K. Power, "Obstacles to decision making in Agile software development teams," *J. Syst. Softw.*, vol. 85, no. 6, pp. 1239–1254, Jun. 2012.

[56] Institute for Disability Research, Policy & Practice, Utah State University, "WAVE Web Accessibility Evaluation Tools," 2001. https://wave.webaim.org/ (retr. Oct., 2023).

[57] T. Boren and J. Ramey, "Thinking aloud: reconciling theory and practice," *IEEE Trans. Prof. Commun.*, vol. 43, no. 3, pp. 261–278, Sep. 2000.

[58] K. S. Fuglerud and D. Sloan, "The Link between Inclusive Design and Innovation: Some Key Elements," in *Human-Computer Interaction. Human-Centred Design Approaches, Methods, Tools, and Environments*, Springer Berlin Heidelberg, 2013, pp. 41–50.

[59] J. Venable, J. Pries-Heje, and R. Baskerville, "FEDS: A Framework for Evaluation in Design Science Research," *European Journal of Information Systems*, vol. 25, no. 1, pp. 77–89, 2016.

[60] C. Power and H. Petrie, "Working With Participants," in *Web Accessibility*, Y. Yesilada and S. Harper, Eds., in Human–Computer Interaction Series. London: Springer, 2019, pp. 153–168.

[61] J. T. Simon-Liedtke, W. K. Bong, T. Schulz, and K. S. Fuglerud, "Remote Evaluation in Universal Design Using Video Conferencing Systems During the COVID-19 Pandemic," in *Universal Access in Human-Computer Interaction. Design Methods and User Experience*, Springer Int. Publishing, 2021, pp. 116–135.

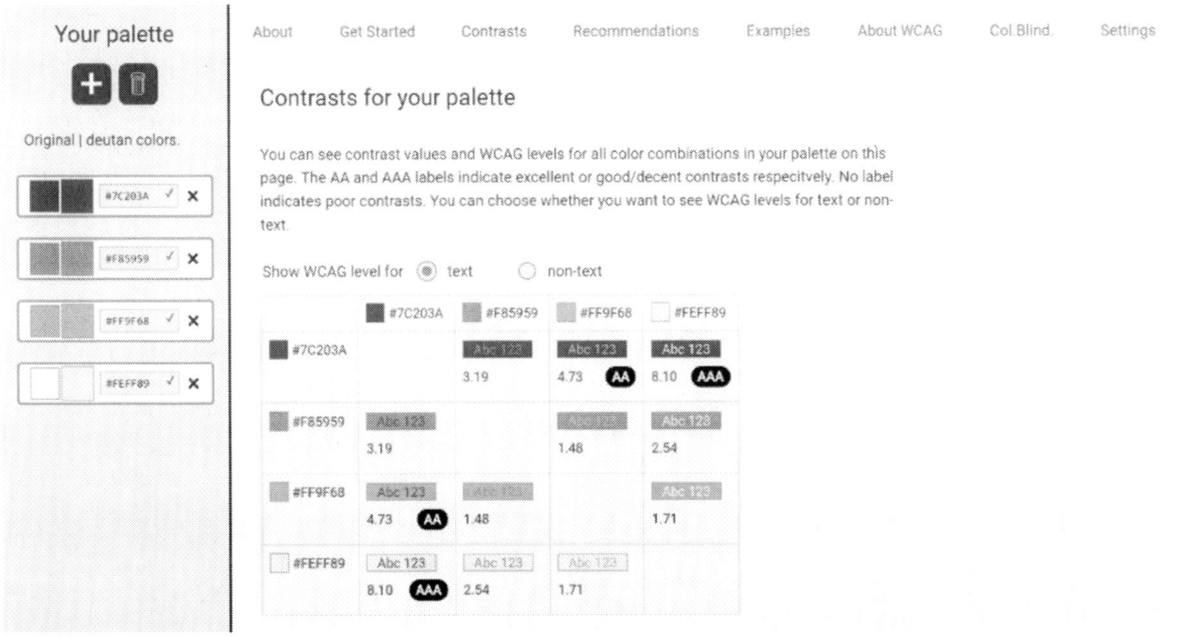

Figure 1. The Color Contrast Checker shows the color contrast calculations and conformance with WCAG 2.1 for textual elements. In this example, colors have been adapted according to an individual with deutan CVD.

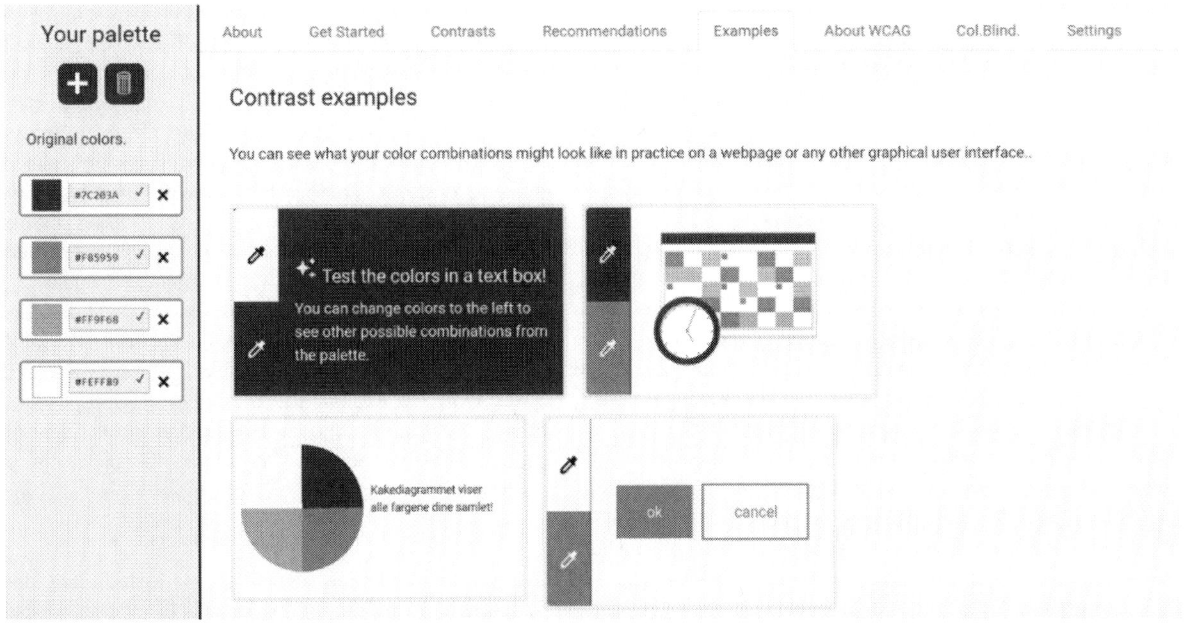

Figure 2. The Color Contrast Example Maker shows how the chosen colors of a palette might graphical elements on a digital interface including text on background, an information graphic, a pie chart, and buttons.

Design and Implementation of Access Control Method Based on Correlation Among Files

Yuki Kodaka
Department of Informatics, The Graduate University for Advanced Studies
Tokyo, Japan
email: y_kodaka@nii.ac.jp

Hirokazu Hasegawa
Center for Strategic Cyber Resilience Research & Development, National Institute of Informatics
Tokyo, Japan
email: hasegawa@nii.ac.jp

Hiroki Takakura
Center for Strategic Cyber Resilience Research & Development, National Institute of Informatics
Tokyo, Japan
email: takakura@nii.ac.jp

***Abstract*—File access control is an effective method for protecting information from unauthorized access both inside and outside an organization. However, conventional methods based on organizational structure have some limitations. Modern business requires flexible access control that reflects the dynamic changes in workflow. Still, it is difficult to achieve the requirement at the same time the prevention of information leakage and destruction due to cyberattacks. Therefore, this paper proposes an access control method based on the correlation among files. The correlation is inferred from users' access behavior within the same group, and access privilege is determined based on the strength of the correlation. This method adapts to changing access needs and prevents unauthorized access by automatically denying access with low file-to-file correlation in a series of accesses. After implementation and verification experiments, it was found that the first determination is the bottleneck of the efficiency of the proposed system. To ensure the feasibility of the proposed system, future work should address this issue.**

***Index Terms*—File access control; Graph theory; Bell-LaPadula model.**

I. INTRODUCTION

File access control has long been used as an effective method of protecting an organization's information assets. It prevents unauthorized accesses by users and minimizes information leakage due to cyber attacks. Various access control methods have been proposed and developed [1]–[3].

However, many of the current methods and operations are not flexible enough. Due to changes in the situations, access control loses accuracy over time [4]. In some cases, policymakers (high-level policy architects) and implementers of policy designed by others are separated. And policies are often managed by several persons rather than a single person [5]. These also make flexible operation difficult.

Strict access control is required, especially in environments where sensitive information is handled. For example, the Bell-LaPadula model [6] was proposed to prevent the leakage of information known only to the supervisor to subordinates. However, in many cases supervisors can write to files that their subordinates can read and write to.

According to Proofpoint report [7], the cost of insider threats has surged from $8.30 million in 2018 to $15.38 million in 2022, an 85% increase. In order to mitigate insider threats, not only technical approaches like access control system, but also non-technical approaches like user behavior analytics are needed [8].

To address these issues, we point out two challenges. One is who and how to determine the need for access. In the proposed method, the determination criteria are based on the user's access behavior. Two is how to assign the necessary access privileges for users. Obviously, the assignment of access privileges should be done with caution. Excessive access privileges increase the risk of leakage or destruction. On the other hand, insufficient access privilege affects the ability of users to perform their operations. As a result, it may undermine the efficiency and productivity of the organization.

To solve these problems, we have proposed an access control method based on the correlation among files [13]. The correlation is inferred from user's access behavior. The method automatically determines whether access is allowed or denied based on the degree of the correlation. It responds to access needs based on changing situations. The system automatically denies accesses with low correlation. It prevents excessive expansion of the access privilege. It is assumed that access by malware is an uncorrelated access. Or, even access by an insider is assumed to be uncorrelated if it is not related to the person's business. These accesses are different from legitimate users. This method can prevent such file accesses. In this paper, we have modified the architecture of our system and performed a brief implementation and verification experiment.

This paper is organized in the following sections. Section II refers to related work to this paper. Section III describes the assumptions of the proposed system and issues in file sharing. After that, we explain the design of the proposed system. Section IV describes the implementation and verification of the proposed system. Section V concludes this paper and presents future work.

II. RELATED WORK

Users are sometimes denied access to files they need, and administrators are required to modify the access control of the files. They might make a misconfiguration at the modification that gives more access privileges than necessary. Xu et al. investigate how and why such problems occur [10]. Although several reasons for misconfiguration are shown, administrators must solve such problems by themselves, and the possibility of misconfiguration and the burden on administrators remains.

Beckerle and Martucci propose the metrics to evaluate and quantify access control rule sets in terms of security and usability [11]. The metrics helps users generate better rule sets. One of the evaluation indicators is the difference between the owner's intention and the rule set. However, the actual method of getting the intention is out of the scope of the paper.

Mazurek et al. propose reactive policy creation in response to user's access request [12]. The experiment involves sharing files on digital devices at home with people, including supervisors and co-workers. If a user tries to access a resource but lacks sufficient privilege, they can use the proposed system to send a request to the resource owner, who can opt to update their policy and allow the access. This method requires the file owner to make determinations for all unauthorized access.

Shalev et al. propose an improved method for containers that allows monitoring and logging of operations by the system administrator [13]. The operations used by system administrators include not only support by internal IT department employees, but also by third parties such as storage service providers and automated management tools used by the IT department. The system administrator is expected to operate based on user requests (tickets in this paper), but there is no mention of whether or not those requests are required.

Desmedt and Shaghaghi propose an access control method that considers three dimensions: subject, object, and operation, rather than the conventional two dimensions of subject and object [14]. These mainly counter internal threats and provide granular access control by controlling operations. It shows how to implement granular access control, but does not mention how to update access privileges once they have been set.

III. Proposed System

In this paper, we proposed a file correlation-based access control method which is modified from our previous works.

A. Assumption

The proposed method assumes an organization consisting of a hierarchical structure as shown in Fig. 1. This paper calls the largest segment of an organization, such as a department in a typical enterprise, a group. Divided units within the group are called subgroups, and further divided units within a subgroup are called subsubgroups. In the example shown in Fig. 1, each department is a group, and each section is a subgroup.

Fig. 2 shows the assumed access control environment. Generally, an Access Control List (ACL) is implemented with coarse-grained, such as per folder, for groups or subgroups. The access privilege under a folder is determined using the information in a user management database, such as Active Directory (AD). If fine-grained access control is to be implemented, it is set by the file owner or the administrator, but their load becomes significant.

In this paper, resources shared within each group are targeted, and resources shared across groups are out of scope.

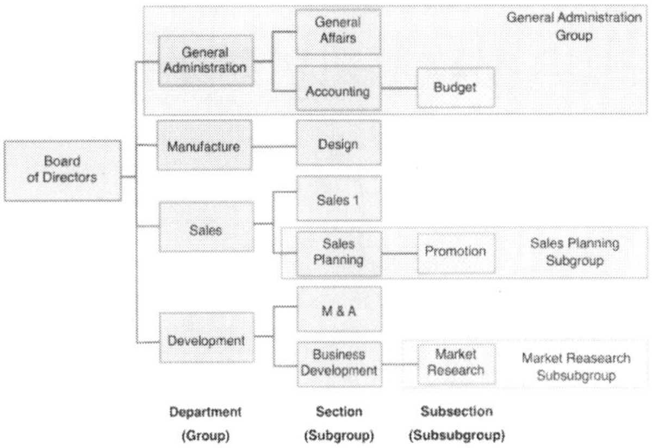

Fig. 1: Example of Organizational Structure

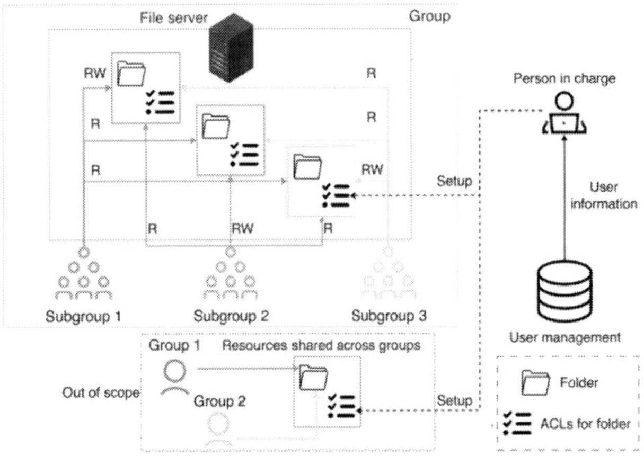

Fig. 2: Assumed Access Control Enviroment

B. Issues in File Sharing

As described in Section I, access control for file sharing involves a risk/benefit tradeoff. Therefore, it is important to balance risks and benefits.

In addition, group or subgroup-based access control alone cannot consider the hierarchical relationship of users in the organization. There is a risk of information leakage through human interaction. It is possible for supervisors to write information in a file that can be read by their subordinates. This could potentially lead to the leakage of information that is known only to the user's rank.

C. Overview of Proposed System

For addressing the file sharing issues, our proposal automatically changes access privileges based on access history for certain period to allow or deny per file, not per folder. It also controls read and write privileges more granularly based on the user's rank. It prevents upper-ranked users from writing confidential information in files that lower-ranked users can read. The proposed method provides hierarchical access control according to the ranks within the organization.

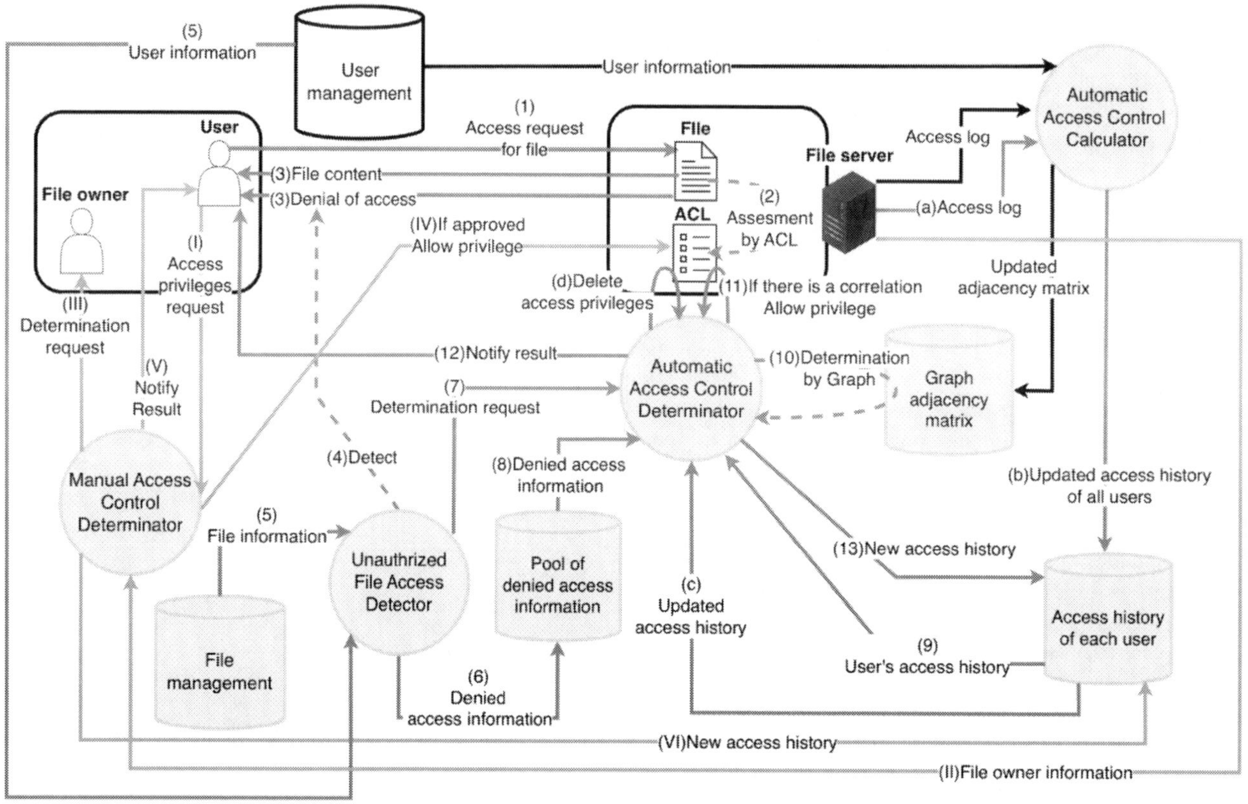

Fig. 3: Architecture of Proposed System (yellow colored)

1) Deletion of unnecessary access privileges: The proposed method records the file access history of each user for a certain period in the past (in this paper, one month). If a user has not accessed a file for this period, the access privilege is considered no longer needed, and it is deleted.

2) Addition of necessary access privileges: When a user tries to access a file without access privileges, such access is denied first. Then the proposed method performs an automatic access determination on the denied access. If the results of the determination show that there is a correlation between the files to which the user has access privilege and the denied files, the ACL is changed to "allowed" to access the file. Even if the access is denied as a result of the automatic access determination, the user can request a manual access control determination if the access is truly necessary. In this paper, it is assumed that manual determination is performed by the file owner.

D. Architecture of Proposed System

Fig. 3 shows the architecture of the proposed system. In Fig. 3, the proposed system is colored yellow. It includes the assumed flow of access determination and the source of information necessary for the determination. The proposed system consists of Automatic Access Control Calculator (AACC), Unauthorized File Access Detector (UFAD), Automatic Access Control Determinator (AACD), Manual Access Control Determinator (MACD), and four databases store the target file information, the denied access information, the access history of each user for a certain period, and the adjacency matrix of the graph. Details will be given later, but the overview of the proposed system process is as follows.

- Deletion of unnecessary access privileges (blue line)
  - (a) AACC receives access logs for a certain period in the past
  - (b) AACC updates access histories of each user
  - (c) AACD receives updated access histories and identifies files that the user has not accessed for a certain period
  - (d) AACD deletes the user's access privileges from ACL for the identified files
- Automatic determination of access privileges (red line)
  - (1) Users access files
  - (2) File server determines whether the access is allowed or denied based on ACL set for each file
  - (3) If allowed, the user gets the file content
    If denied, the user is notified of denial
  - (4) UFAD detects the denied log
  - (5) UFAD fetches the target file information from the database for file management and the target user information from the database for user management

(6) If the content of log matches the target files and the users, UFAD store it in the database that stores the denied access information
(7) UFAD requests an access determination to AACD
(8) AACD requests the access information from the database that stores the denied access information
(9) AACD requests the user's access history from the database that stores the access history of each user for a certain period
(10) AACD performs the access determination using graphs
(11) If there is a correlation, AACD allow access privilege to user
(12) AACD notifies the result to the user
(13) AACD adds the newly allowed file to the user's access history

- Manual determination of access privileges (green line)

(I) MACD receives determination requests from users
(II) MACD requests the file owner information from File server
(III) MACD requests the file owner to determine whether the access is allowed or not
(IV) If approved, MACD allow the access privilege to the user
(V) MACD notifies the result to the user
(VI) MACD adds the newly allowed file to the user's access history

The functions of AACC and AACD are described below.

E. Automatic Access Control Calculator (AACC)

Calculator creates graphs utilizing graph theory for correlation determination. The graph infers the correlation among files based on the user's access behavior.

The calculation procedure is as follows. Data is access logs for a certain period, which is the past month (the past 30 days) in this paper.

a) Extract specific information from access logs: Specific information in the access log is recorded as the access log used in the calculation. The specific information is "Timestamp", "AccessType"(Read or Write), "UserName", "Filename". The extracted access logs are sorted by username and time.

b) Categorize access logs by user rank and access type: Extracted access logs are categorized by user rank and access type. Ranks are assumed to be hierarchical. For example, from the top, director, manager, section chief, member. For each user rank, two access logs are categorized. One is the access log of the "Read" access type for users below the same rank. The other is the access log of the "Write" access type for users in the same rank.

c) Create graphs from access history: An example graph is shown in Fig. 4. The graph consists of nodes (V_1, V_2, V_3) and links between nodes (L_{1-2}, L_{2-3}, L_{1-3}). In the graph, nodes represent files. Links represent the correlations among files. The graph is assumed to be undirected. The order of accesses, A-B and B-A are counted as the same.

The graph is calculated using an adjacency matrix. Adjacency means that node i and node j are adjacent to link $i-j$

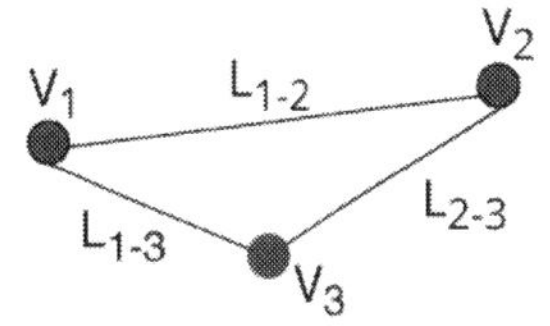

Fig. 4: Example of Graph

in the graph. An adjacency matrix is a square matrix used to represent a finite graph. The elements of this matrix indicate if a pair of nodes is adjacent or not in the graph. If so, it indicates the weights of the links between adjacent nodes. An example of an adjacency matrix is shown in Table I.

TABLE I: EXAMPLE OF ADJACENCY MATRIX

	FileA	FileB	FileC	FileD
FileA	0	3	0	1
FileB	3	0	1	5
FileC	0	1	0	1
FileD	1	5	1	0

The following procedure is used to calculate link weights.

a. Get a list of files by rank and access type from categorized logs
b. Determine the size of the adjacency matrix from the list
c. Create adjacency matrix initialized to 0
d. Calculate weights of links from access logs
Add link weights between consecutive files in the categorized access history, if the same user accesses different files within a certain period of time (one hour in this case). Furthermore, we add the time inclination shown in (1) based on the timestamp:

$$1-\left(\frac{D}{D_{\max}}\right)^{n} \tag{1}$$

where D is the number of days elapsed from the most recent day, $D_{\max}$ is the number of calculation days, and n is an adjustment parameter.
e. Normalize weights of links
Let A be the adjacency matrix before normalization, and $S(n)$ be the total weight of the links connected to each node n. Normalization is performed as shown in (2):

$$B(i,j)=\frac{A(i,j)}{S(i)}+\frac{A(j,i)}{S(j)} \tag{2}$$

where $B(i,j)$ is the element at the ith row and jth column of the adjacency matrix after normalization, $A(i,j)$ is the element at the ith row and jth column of the adjacency matrix before normalization, $S(i)$ is the total weight of the links connected to node i, $A(j,i)$ is the element at the jth row and ith column of the adjacency matrix before normalization, and $S(j)$ is the total weight of the links connected to node j. Also, round off to the second decimal place.

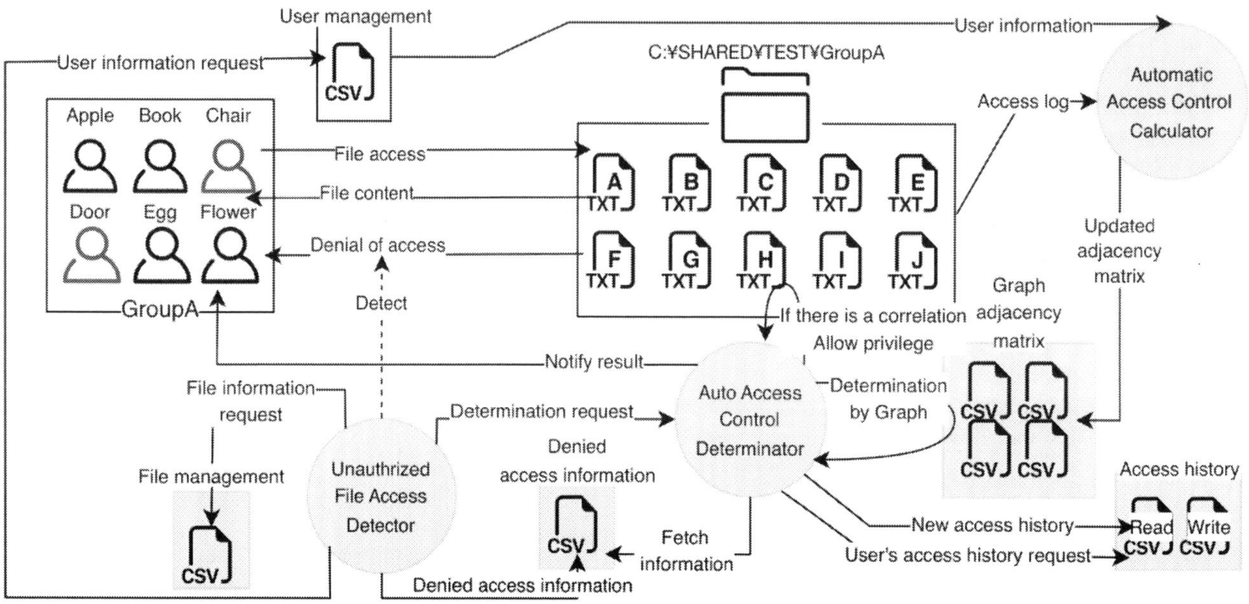

Fig. 5: Implementation of Proposed System (yellow colored)

Table II shows the results of the above calculation steps using Table I as an example.

TABLE II: EXAMPLE OF NOMALIZED ADJACENCY MATRIX

	FileA	FileB	FileC	FileD
FileA	0.00	1.08	0.00	0.39
FileB	1.08	0.00	0.61	1.27
FileC	0.00	0.61	0.00	0.64
FileD	0.39	1.27	0.64	0.00

F. Automatic Access Control Determinator (AACD)

The determination method is as follows. If the elements of the adjacency matrix exceed a certain threshold, it is considered correlated. The formula is shown in (3). Here, as an example, the threshold is set at 0.8 or higher.

$$\text{Matrix}(\text{File}_{\text{old}}, \text{File}_{\text{new}}) \geq 0.8 \tag{3}$$

where $\text{Matrix}(\text{File}_{\text{old}}, \text{File}_{\text{new}})$ is the correlation between files, File_{new} is the new file to be accessed by user u, and File_{old} is the file already accessed by user u.

IV. Implementation and Evaluation Experiment of Proposed System

A. Implementation of Proposed System

The proposed system was implemented in a brief experimental environment. As the hardware, we used Intel® NUC 8 Pro Kit (NUC8v7PNH). As OS, we used Windows 11 pro edition. The implemented environment is shown in Fig. 5. We set up 6 users, Apple, Book, Chair, Door, Egg, and Flower as general users. We set up 3 users, AACC, UFAD, and AACD, to take on the roles of the proposed system. We created 10 files named A, B, C, D, E, F, G, H, I, J in a folder with the path "C:¥SHARED¥TEST¥GroupA".

The access log in Windows is "Security" in "Windows Log" (hereinafter referred to as Windows Security Log).

The proposed method needs to notify the determination results to the user. There are several possible ways to notify such as e-mail, text message, phone call or other means. We made notification possible by running the program on both the receiver and the sender of the message during experiment.

The following is a description of the setting for the system and the implementation of each element of the system.

1) Setting File Privileges for Users: The privileges allowed to each user for each file are shown in Table III.

2) Automatic Access Control Calculator (AACC): The graph calculation was set up in two stages. The first stage is calculated at 1:00 a.m. daily using data from 30 days prior to the previous day. The second stage is calculated every hour during business hours, using data up to the present time of the day. After the second stage of calculation, the graphs from the first and second stages were combined and normalized. This is because graph calculation takes a lot of time.

The calculation used Event ID 4663 (An attempt was made to access an object.) from Windows Security Log.

3) Unauthorized File Access Detector (UFAD): Using "Task Scheduler", UFAD was triggered by a log that access to a file by the user was denied. The log was Event ID 4656(A handle to an object was requested.) from Windows Security Log. Only failures were collected.

As well as way of notification to the user, UFAD and AACD were contacted by executing the program.

TABLE III: FILE ACCESS PRIVILEGES FOR EACH FILE ALLOWED TO EACH USER

User	FileA	FileB	FileC	FileD	FileE	FileF	FileG	FileH	FileI	FileJ
A	R/W	R/W		R/W	R/W	R	R/W	R/W		R/W
B		R	R/W	R/W	R/W	R/W	R/W	R		R/W
C	R/W	R/W		R/W			R/W	R/W	R/W	R/W
D	R/W	R/W	R	R/W	R/W	R/W			R/W	R/W
E	R/W				R/W	R	R/W	R/W	R/W	R/W
F	R/W	R/W		R/W	R/W	R/W	R/W		R/W	R

UFAD recorded the output access log. In this evaluation, the last 10 access logs were recorded. It also fetched the latest access logs before outputting the denied log. In this case, the latest 10 access logs were fetched. If there is no output in spite of the target log, the logs will be output along with the denied log.

4) Automatic Access Control Determinator (AACD): AACD was listening with a request from UFAD. When it is received, AACD execute processes. The determination was repeated as long as there was information in Pool of denied access information. When there was no more information, AACD waited again in the listening state.

AACD recorded the history of past determinations. In this paper, it was recorded for the past 10 minutes. If the same file was accessed and rejected within 10 minutes, the determination was made only once, and the rest of the accesses were skipped without determination.

5) CSV File for User management: This file recorded information in the following three columns.

- UserName (Apple, Book, Chair, Door, Egg, Flower)
- Rank (Chair and Door is rank 2, the rest is rank 1)
- Group (all users are Group A)

6) CSV File that stores target file information: This file stored a list of target files (A, B, C, D, E, F, G, H, I, J).

7) CSV File for Pool of denied access information: This file stored the denied log information Timestamp, AccessType, UserName, FileName.

8) CSV Files for Graph adjacency matrix: These files stored the adjacency matrices of the graph calculated by AACC.

9) CSV Files for Access history of each user: These files stored the access history of each user for each access type.

B. Evaluation Experiment of Proposed System

In this section, the efficiency of the proposed system is verified as an evaluation experiment. As a measure, the response time was tested. Methodology are shown below.

1) Methodology: Two types of experiments were conducted. First, one to six users simultaneously accessed an unauthorized file. We verified the change in processing time due to the change in the number of users. Second, one to six users accessed unauthorized files at regular intervals from the previous user. In this experiment, the next user accessed the file at 5-second delays. We verified the change in processing time when several users accessed at regular intervals. The number of users was increased from one to six, and each was performed five times.

Access is done by executing a script at a specific time using the task scheduler. The script indicates the name of the file to be accessed. It is a file to which each user does not have privileges. The script for user Apple describes file C. Similarly, the script for Book describes file A, the one for Chair describes file C, the one for D describes file G, the one for Egg describes file B, and the one for F describes file C. The time from when the script is executed to when the determination results are notified to the user is measured as the response time.

TABLE IV shows the data sets used in the graph calculations for the determination. Since the data set up to the previous day has 404 rows, only a summary of the data set is shown. Up to the day is without parentheses and the day is surrounded by parentheses.

2) Results: Experimental results for simultaneous access and access with 5-second delays are shown in Table V. Number of user columns indicates the number of accessing users and the initial letter of the accessing user. For example, 1(A) indicates that a single user named Apple accessed an unauthorized file. The columns 1st trial through 5th trial show the response times for each number of users. If the number of users is 2 or more, the latest response time among users is noted. 5-trial average column shows the average of five trials for the same number of users. Average per user is calculated by dividing the 5-trial average by the number of users in the corresponding row. In this paper, a case is defined as the number of times a determination is made to allow or deny access.

C. Discussion

Table V shows that less determination time is required after the second case. Comparing the simultaneous access of 1(A) and 2(A/B), for example, the average response times per user become 2.8 seconds shorter. This experiment shows that average response time per user decreases as the number of users increases. This indicates the efficiency of the system improves after the second case. However, this also shows the first determination is the bottleneck of the efficiency of the proposed system.

Table V shows that it takes approximately the same amount of time to make a determination from the first case to the sixth case. From this result, it can be inferred that one cycle of determination is completed after the 5-second delays. However, the first case took a little bit longer to determine as same as the result of simultaneous access.

Limitation: In this paper, the proposed system has only been able to verify the situation with six users. However,

TABLE IV: DATA SET SUMMARY FOR GRAPH CALCULATION

User	FileA		FileB		FileC		FileD		FileE		FileF		FileG		FileH		FileI		FileJ	
	R	W	R	W	R	W	R	W	R	W	R	W	R	W	R	W	R	W	R	W
A	7(2)	5(1)	2(1)	0(0)	0(0)	0(0)	3(2)	1(0)	3(1)	0(1)	4(1)	0(1)	4(1)	4(1)	6(0)	4(0)	0(0)	0(0)	0(0)	0(0)
B	0(0)	0(0)	6(0)	0(0)	6(2)	5(1)	6(2)	6(2)	7(1)	6(1)	9(0)	8(0)	7(0)	7(0)	10(1)	0(0)	4(0)	0(0)	0(0)	0(0)
C	5(1)	5(0)	3(1)	1(1)	0(0)	0(0)	8(1)	8(1)	0(0)	0(0)	0(0)	0(0)	2(1)	1(0)	5(1)	5(1)	4(1)	3(1)	0(0)	0(0)
D	2(1)	1(1)	7(0)	7(0)	6(2)	0(0)	10(2)	7(0)	5(2)	4(1)	10(1)	7(0)	0(0)	0(0)	0(0)	0(0)	6(0)	6(0)	0(0)	0(0)
E	9(2)	6(1)	0(0)	0(0)	0(0)	0(0)	0(0)	0(0)	7(2)	7(1)	6(0)	0(0)	14(1)	13(1)	4(0)	3(0)	9(2)	7(0)	0(0)	0(0)
F	6(1)	3(1)	5(1)	5(1)	0(0)	0(0)	5(1)	4(1)	6(1)	4(0)	5(0)	4(2)	5(2)	4(0)	0(0)	0(0)	3(1)	2(0)	0(0)	0(0)

TABLE V: EXPERIMENTAL RESULTS: RESPONSE TIME

	Response Time (s)						
Number of User	1st trial	2nd trial	3rd trial	4th trial	5th trial	5-Trial Average	Average per User
Simultaneous Access							
1(A)	6.848931	6.220957	6.619395	5.960975	6.688213	6.4676942	6.4676942
2(A/B)	7.961715	7.919432	6.505913	6.712306	7.820161	7.3839054	3.6919527
3(A/B/C)	7.905125	9.562813	7.342283	9.489325	8.582019	8.576313	2.858771
4(A/B/C/D)	8.611524	8.262025	8.348701	8.788181	8.162078	8.4345018	2.10862545
5(A/B/C/D/E)	8.791258	8.952904	9.555604	9.10041	9.461176	9.1722704	1.83445408
6(A/B/C/D/E/F)	11.200058	12.773345	9.905579	12.539686	11.843131	11.6523598	1.942059967
Access with 5s Delay							
1(A)	6.833407	6.63882	6.87596	8.329752	6.656232	7.0668342	7.0668342
2(A/B)	11.263513	11.485306	11.795818	12.237531	11.947039	11.7458414	5.8729207
3(A/B/C)	16.516348	16.310971	15.935014	16.120887	16.474012	16.2714464	5.423815467
4(A/B/C/D)	23.57539	21.183435	22.799851	23.252516	21.822607	22.5267598	5.63168995
5(A/B/C/D/E)	27.278206	27.148512	26.288184	26.199733	27.018846	26.7866962	5.35733924
6(A/B/C/D/E/F)	31.92014	31.139518	31.577622	31.936971	31.464357	31.6077216	5.2679536

it cannot guarantee scalability beyond that. For example, when the number of users reaches 10 or 50 or more, There is a possibility of leaks or duplicates in UFAD detection and processing. Considering the bottleneck of the first case determination, it is necessary to guarantee the scalability and efficiency of UFAD in the future.

In this proposal, the determination is made based on the past access history. It does not solve the problem of how to set the initial settings for newly created files or when a user's department changes. These issues need to be addressed as well.

V. Conclusion and Future Work

In this paper, we propose an access control method that responds to changing situations. It automatically blocks continuous access to files with low correlation. It enables automatic determination and reduces administrative burdens. If there is no or low correlation, detailed manual determination prevents unauthorized access.

After implementation and verification experiments, It was found that the first determination is the bottleneck of the efficiency of the proposed system. In addition, the scalability of it has not yet been verified. To make the system feasible, future work should address the issue of efficiency and scalability. It is also essential to have an adequate data set for verification.

There are some thresholds that need to be set. For example, the extraction period of access logs is set to the past month. The threshold for correlation is a matrix element greater than or equal to 0.8. These thresholds are tentative. They are subject to change depending on the target organization and cyber attack stages. Detailed discussion is required to set them.

Acknowledgment

This work was partially supported by JSPS KAKENHI Grant Number JP19K20268.

References

[1] P. Samarati and S. C. Vimercati, "Access control: policies, models, and mechanisms," Foundations of Security Analysis and Design, R. Focardi, R. Gorrieri, ed., Springer, pp.137-196, 2001.

[2] D. F. Ferraiolo and D. R. Kuhn, "Role-based access control," 15th National Computer Security Conference, pp.554-563, 1992.

[3] V.C. Hu, D. Ferraiolo, R. Kuhn, A. Schnitzer, K. Sandlin, R. Miller, and K. Scarfone, "Guide to Attribute Based Access Control (ABAC) Definition and Considerations," U.S. Department of Commerce, 2014.

[4] H. Xia, M. Dawande, and V. Mookerjee, "Role refinement in access control: model and analysis," INFORMS Journal on Computing vol.26, no.4, pp. 866-884, 2014.

[5] L. Bauer, L. F. Cranor, R. W. Reeder, M. K. Reiter, and K. Vaniea, "Real life challenges in access-control management," in Proceedings of the CHI Conference on Human Factors in Computing Systems, Association for Computing Machinery, pp. 899-908, 2009.

[6] D. E. Bell and L. J. LaPadula, "Secure computer systems: mathematical foundation report ESD-TR-73-275," MITRE Corp., 1973.

[7] Ponemon Institute, "2022 cost of insider threats global report," Proofpoint, 2022.

[8] D. Tsiostas et al., "The insider threat: reasons, effects and mitigation techniques," in 24th Pan-Hellennic Conference on Informatics, pp.340-345, 2020.

[9] Y. Kodaka, H. Hasegawa, and H. Takakura, "A proposal for access control method based on file relation inference from users behavior(in Japanese)," IEICE Technical Report vol.123, no.86, pp. 40-47, 2023.

[10] T. Xu, H. M. Naing, L. Le and Y. Zho, "How do system system administrators resolve access-denied issues in the real world?," in Proceedings of the CHI Conference on Human Factors in Computing Systems, pp. 348-361, 2017.

[11] M. Beckerle and L. A. Martucci, "Formal definitions for usable access control rule sets from goals to metrics," in Proceendings of the Ninth Symposium on Usable Privacy and Security, pp. 1-11, 2013.

[12] M. L. Mazurek et al., "Exploring reactive access control," in Proceedings of the CHI Conference on Human Factors in Computing Systems, Association for Computing Machinery, pp. 2085-2094, 2011.

[13] N. Shalev, I. Keidar, Y. Weinsberg, Y. Moatti, and E. B. Yehuda, "WatchIT: who watches your IT guy," in Proceedings 26th Symposium on Operating Systems Principles, pp.515-530, 2017.

[14] Y. Desmedt and A. Shaghaghi, "Function-based access control (FBAC) from access control matrix to access control tensor," From Database to Cyber Security, vol 11170, pp.143-165, 2018.

Towards a Minimalistic Stress Classification Method based on HRV

Roswitha Duwenbeck
Systeme der Medizintechnik
Universität Duisburg-Essen
Duisburg, Germany
E-mail: roswitha.kressner@uni-due.de

Elsa Andrea Kirchner
Systeme der Medizintechnik & Robotics Innovation Center
Universität Duisburg-Essen & German Research Center for Artificial Intelligence
Duisburg, Germany & Bremen, Germany
E-mail: elsa.kirchner@uni-due.de & elsa.kirchner@dfki.de

***Abstract*—Stress is a feeling of emotional and physical tension, that poses as a risk factor in many diseases, for example the nervous, musculoskeletal, cardiovascular or gastrointestinal system. Fast and easy detection could be a first step in order to help people manage their stress-levels. This paper depicts an ongoing work in the domain of stress prediction with Heart Rate Variability related features by classifying two different levels on the Stress-Predict Dataset. The performance of different classifiers was tested with Leave-One-Subject-Out Cross Validation and compared to each other. The best performance was reached with the Aggregated Mondrian Forest Classifier and a mean balanced accuracy of 97.87%.**

***Index Terms*—Heart rate variability; stress prediction; Machine Learning.**

I. Introduction

There are manifold definitions of stress in humans. Maybe the most generic and well-known definition is the one by Hans Seyle, stating that "Stress is the non-specific response of the body to any demand" [1]. This demands can be of physical or psychological nature [2] [3]. Especially psychological stress, which can be defined as "stress that occurs when an individual perceives that environmental demands tax or exceed his or her adaptive capacity" [4], is a topic of great interest in healthcare. While both stressors, psychological and physiological, are of very different nature and by thus, have different effects on the body, they also share a subset of comparable effects on the body [5]. Not only do they cause similar subjective and hormonal responses, it has also been suggested that they also share common neural substrates [6]. Long term effects of both stressors seem to show more differing symptoms on the body. While effects of too much psychological stress are well documented and mostly about harm of the nervous, musculoskeletal, respiratory, cardiovascular, gastrointestinal, reproductive, and other systems [7], the influences of physical stress are not so clearly outlined. Nevertheless there are studies, which observed effects like increased luminal permeability [8] or differences in corticosterone serum-levels [9]. Short term effects of both stressors tend to be more similar, but still differing. Both stress types cause a rise in physiological parameters like Heart Rate (HR), Breathing Rate or Oxygen Consumption, but differ in severity. Like this, the Oxygen Consumption and Breathing Rate is higher in physical than in psychological stress. The HR, on the other hand, is higher in psychological stress. [11]

The possible damages outline the necessity of a reliable stress detection method. Can et al. state, that researchers found out that stress should be handled when the symptoms first come out to avoid the long-term consequences [12]. This can be important in different settings like the workplace, traffic and generally in healthcare.

Because of its importance, automated stress detection is not a new topic in the area of Machine Learning (ML). There are already many different approaches in terms of using psychophysiological signals, selected features, and machine or deep learning methods. Can et al. used different ML methods, such as Linear-Discriminant Analysis (LDA), Support-Vector-Machine (SVM), k-nearest-Neighbors (kNN), Logistic Regression (LR), Random Forest (RF) and Multi Layer Perceptron (MLP) to detect 3 psychological stress-classes with the help of HR, Electrodermal Activity (EDA), Inter Beat Intervalls (IBI), Skin Temperature (ST) and Acceleration [12]. Costin et al. used Heart Rate Vsariability (HRV) related features from the Electrocardiogram (ECG), to train a Minimum Distance Classifier (MDC) and detect three psychological levels of stress [13]. Garg et al. used ECG, body temperature (TEMP), Respiration (RESP), Electromyogram (EMG), and EDA to classify two or three physiological conditions - neutral (baseline), psychological stress and neutral (baseline), psychological stress, amusement with the help of kNN, LDA, RF, AdaBoost (AB), and SVM [14]. Their best

TABLE I
SELECTION OF PAPERS DETECTING STRESS WITH ML.

Paper	Details		
	Signals	***ML-Method***	***Best Results***
[12]	HR, EDA, IBI, ST, Acceleration	LDA, SVM, kNN, LR, RF, MLP	Accuracy of 97.92%
[13]	ECG (HRV)	Minimum Distance Classifier	Accuracy of 89.36%
[14]	ECG, TEMP, RESP EMG, EDA	kNN, LDA, RF AB, SVM	Accuracy of 84.17%
[15]	ECG (HRV)	kNN, SVM, MLP RF, GB	F1 of 79%
[16]	ECG(HRV)	SVM, MLP, IBK DT, LDA	Accuracy of 94%

classification results were reached in the binary classification task with RF and 84.17% Accuracy, while in the three class problem they reached 67.56% Accuracy, also with RF [14]. Dalmeida et al. classified psychological stress in two classes with ECG-derived HRV-Features and different ML methods such as kNN, SVM, MLP, RF and Gradient Boosting (GB) [15]. Castaldo et al. classified two classes of psychological stress by using ECG derived HRV-Features with SVM, MLP, Neighbor Search (IBK), DT and LDA [16].

It becomes clear that the classification of stress can be done with the help of many psychophysiological features. And although using HRV traits alone does not appear to be as accurate as combining multiple signals, it is nevertheless an interesting approach for minimal applications. This paper can be seen as a starting point in creating a minimalist stress classification method, which is robust and practical for the in use real world scenarios. This paper is structured as follows: In Section II the used materials, the Dataset and it´s preparation, the chosen features and the ML methods and validation procedure are explained. Section III describes the validation results, while Section IV draws conclusions out of the results and lists future plans for the project, as this paper depicts an ongoing work.

II. Methods and Materials

The Stress-Predict Dataset (SPD) was used to train different ML classifiers in a binary classification task, classifying a stressed and a rest state. To make training possible, steps such as preparing the dataset, feature extraction and choosing the ML methods had to be done. These steps, and also the used materials, are described in here.

A. The Dataset

The relatively new SPD from Iqbal et al. consists of bio-signals from 35 participants, 25 men and 10 women. Stressors were forced Hyperventilation, the Trier Social Stress Test and Stroop Color Test. An E4 watch from Empatica was used to measure individual physiological changes based on PPG. The signal was filtered to get a clean Blood Volume Pulse (BVP), which was used to obtain the HR, IBI and Respiration Rate (RESP) by an estimation algorithm. [18]
To the authors knowledge this dataset was not used before to detect stress from HRV features.

B. Preparation of the dataset and preprocessing

Because the dataset was not mainly composed to detect stress from HRV-Features, but from HR and RESP, an own assignment of labels, timestamps and physiological parameters had to be done. Iqbal et al. original distribute one "processed" data folder and one with raw data. The first folder contains a list, with merged patient label, HR, RESP a stress-label and a timestamp for every second, given in ms with one decimal place. The raw data contains separate lists of the physiological signals with the passed time since the start, given in ms with 6 decimal places, and the starting time of the experiment in ms as header, for every subject.

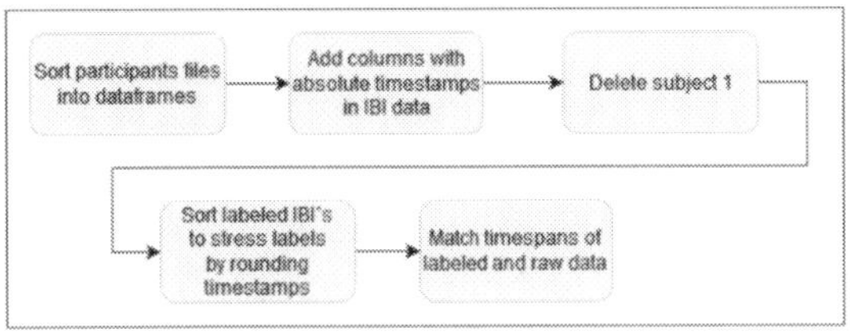

Fig. 1. Dataset preparation

It was important to assign labels to the IBIs. At the start, subject data was sorted into dataframes. It could be seen, that subject one has no matching labeled data, so it was deleted. The raw condition of the IBI dataframes just contains the IBI´s and the passed time in ms from the respective starting time. To facilitate sorting labels to IBI´s, a new column for the absolute passed time since start, as given in the processed filed, was added. The starting time was given in the raw header file, by this the columns could be filled with an iterative addition of the start time and the passed time. Also, the processed data is time-wise longer than the raw IBI data. Processed data which could not be associated to any IBI data was therefore deleted. From there on, the labeled data was sorted to each IBI by rounding the IBI-times: Each raw timestamp, given in ms, was rounded to match a processed timestamp given in seconds, and by this sorted to one Stress-Label. The process can be seen in Figure 1. In a next step, the IBI-signals were windowed into 60-sec-windows. Since the windowing sometimes produced windows with two different labels, the window was labeled according to the majority of stress-labels. Windows with less than 30 IBI's were sorted out as it would not be physiological.

C. HRV Features

To extract HRV Features from IBI-signals the pyhrv-toolkit was used [17]. The chosen features can be sorted in time- and frequency-domain-, but also nonlinear features.

1) Time Domain Features: Time domain features included "NN" Parameters, which denote the time between two consecutive R-peaks in an ECG signal [19]. Different statistical features like the NN-Counter, mean, minimum and maximum of the time window and differences in the time window were taken. Furthermore the standard deviation of the NN´s was taken, the standard deviation (SD) of the average NN, the root mean square of successive differences, the number of pairs of successive NN´s that differ by more than 50 ms (nn50) and 20ms (nn20), the proportion of NN50 and NN20 divided by total number of NN´s (pNN50 and pNN20).

2) Frequency Domain Features: To obtain frequency domain features, Welch´s Power Spectral Density was used. For the classification task, the absolute powers of the very low (0.00Hz - 0.04Hz), low (0.04Hz - 0.15Hz) and high frequency (0.15Hz - 0.40Hz) band was used. Also, the total power of all frequency bands and the ratio of the power of the low and high frequency bands.

TABLE II
TESTED ML METHODS

Classifier	Parameters
Dummy Classifier	strategy = most frequent
Multi Layer Perceptron	max_iter=45, hidden_layers=45, 20, batch_size=15
Passive Aggressive Classifier	C=0.0, fit_intercept=False, early_stop=True, max_iter=50
SGD Classifier	penalty='l2', alpha=0.01, max_iter=100, eta0=0.1, epsilon=0.01, early_stop=True
Support Vector Machine One Vs. Rest	C=100.0, degree=10
Gaussian Naive Bayes	All standard
Decision Tree	criterion=entropy
Random Forest	All standard
Support Vector Machine One Vs. One	C=100.0, degree=10
Hoeffding Adaptive Tree	grace_period=100, delta=1e-5, seed=0 leaf_prediction='nb', nb_threshold=10
Hoeffding Tree	grace_period=100, delta=1e-5, binary_split=True
Aggregated Mondrian Forest	n_estimators=5, seed=45
Adaptive Random Forest	n_models=7, seed=45

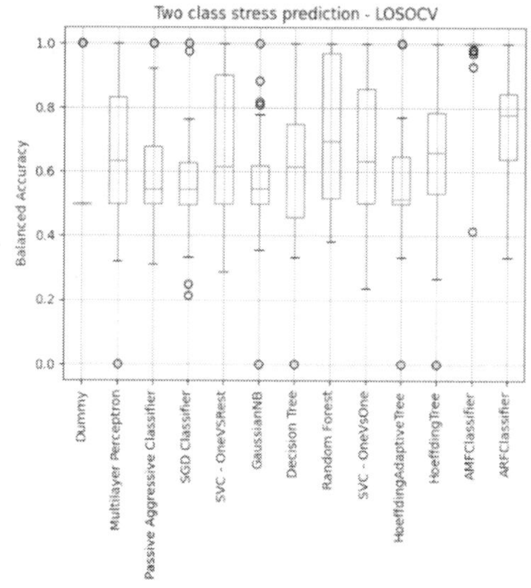

Fig. 2. Boxplots of balanced accuracies

3) Nonlinear Features: Chosen nonlinear features were SD1 and SD2, which are the SD of the data series along the minor axis and the major axis of the Poincaré-Plot.

D. ML-Methods and Learning

A variety of ML methods were used to find the best method for this use case. Used classifiers with their parameters can be found in Table II. The first nine classifiers are from scikit-learn [20], while the last four were taken from river [21]. There was no Hyperparameter Tuning, parameters were rather chosen by experience or trial runs. To gain a better understanding of the robustness of each classifier with respect to completely unseen data, a Leave-One-Subject-Out Cross Validation (LOSOCV) algorithm was written. This own implementation was mainly necessary, because the online learning behaviour of the classifiers based on the river library is hardly compatible with scikit-learn. Before training and testing the data was artificially balanced with Synthetic Minority Oversampling and scaled with a Standard-Scaler.

III. EXPERIMENTAL RESULTS

Boxplots of the mean balanced accuracies are shown in Figure 2. The x-axis shows used classifiers, the y-axis shows the mean balanced accuracies across all subjects. Detailed results can be seen in Table III. Listed are the means over all subjects. It is obvious, that the Aggregated Mondrian Forest (AMF) outperforms the other classifiers by far and keeps up with the state of the art seen in [16].

Lessons learned are that ensemble methods, like RF or the SVM-Methods performed better than non-ensemble learners. Also, non-linear methods, like the Ensemble-SVM´s with radial basis function-kernels, outperformed linear methods, like the Passive Aggressive- or SGD-Classifier. It is generally known that both methods perform better on a large number of samples and/or characteristics, so this is not surprising. The Dummy Classifier performed as expected with one outlier being a subject that only expresses the state of no stress, because of the out-sorting of windows with less than 30 IBI´s.

IV. CONCLUSION AND FUTURE WORK

The AMF was able to classify the two states "No Stress" and "Stress" with nearly perfect results. Why exactly this classifier outperforms others by a large margin, has to be investigated. To this moment at least incorrect infusion of the output label in the testing data has been ruled out. Because (A)MFs are relatively new [21], there is not much literature about best use cases or pro and contra. Generally, MFs seem to prefer sparse feature spaces [21], which is the case here. The authors hope to generate an answer about reasons for the excellent performance in the near future.

To receive an answer regarding the unusal good performance of the AMFs, a future task would be to test the same classifiers on a different dataset. Currently, efforts are being made to use "MIT DriveDB" [23] to gain more knowledge about AMFs, while also testing classification accuracy with more than two classes. The aim of more than two stress classes is to split up them in various intermediate stress levels, as it better fits real-world applications.

An additional goal is to investigate more than one time window, because different window lengths seem to result in varying accuracies. At the moment, 5 minute windows still seem to be the recommended ones [24]. In addition, the use of an individual baseline could be useful: Since each person has slightly different characteristics of HRV related to factors such as age, health, etc., it could lead to better comparability. Furthermore a classical Hyperparameter Tuning could lead to better results for all methods. Finally, as in [7], a long-term goal would be to distinguish between the causes for stress, psychological or physical, to gain better insight into the causes and possible effects of stressful events.

ACKNOWLEDGMENTS

This work is funded by the Federal Ministry for Economic Affairs and Climate Action and the German Aerospace Center (DLR) [grant number 50RP2260A].

APPENDIX

TABLE III
CLASSIFICATION PERFORMANCES (LOSOCV)

Classifier	Performance
Dummy Classifier	Accuracy: 62.65%
	Balanced Accuracy: 52.94%
	Precision: 34.26%
	Recall: 52.94%
Multi Layer Perceptron	Accuracy: 65.62%
	Balanced Accuracy: 65.63%
	Precision: 63.30%
	Recall: 64.30%
Passive Aggressive Classifier	Accuracy: 59.62%
	Balanced Accuracy: 59.53%
	Precision: 56.02%
	Recall: 57.97%
SGD Classifier	Accuracy: 55.03%
	Balanced Accuracy: 56.20%
	Precision: 53.43%
	Recall: 55.71%
Support Vector Machine One Vs. Rest	Accuracy: 67.00%
	Balanced Accuracy: 66.42%
	Precision: 63.13%
	Recall: 65.08%
Gaussian Naive Bayes	Accuracy: 53.55%
	Balanced Accuracy: 56.58%
	Precision: 50.24%
	Recall: 55.38%
Decision Tree	Accuracy: 63.96%
	Balanced Accuracy: 63.13%
	Precision: 63.15%
	Recall: 62.06%
Random Forest	Accuracy: 71.67%
	Balanced Accuracy: 71.96%
	Precision: 69.68%
	Recall: 71.96%
Support Vector Machine One Vs. One	Accuracy: 67.21%
	Balanced Accuracy: 67.21%
	Precision: 63.54%
	Recall: 65.43%
Hoeffding Adaptive Tree	Accuracy: 53.81%
	Balanced Accuracy: 57.37%
	Precision: 48.34%
	Recall: 56.88%
Hoeffding Tree	Accuracy: 61.27%
	Balanced Accuracy: 65.12%
	Precision: 62.61%
	Recall: 63.30%
Aggregated Mondrian Forest	Accuracy: **97.79%**
	Balanced Accuracy: **97.87%**
	Precision: **97.50%**
	Recall: **97.87%**
Adaptive Random Forest	Accuracy: 69.58%
	Balanced Accuracy: 73.89%
	Precision: 68.12%
	Recall: 72.19%

REFERENCES

[1] H. Seyle, Stress in health and disease, Butterworth-Heinemann, p. 15, 2013.

[2] Y. Li, J. Qin, J. Yan, N. Zhang, Y. Xu, Y. Zhu, L. Sheng, X. Zhu and S. Ju, "Differences of physical vs. psychological stress: evidences from glucocorticoid receptor expression, hippocampal subfields injury, and behavioral abnormalities," Brain Imaging And Behavior, vol. 13, pp. 1780-1788, 2019.

[3] S. Rao, R. Hatfield, J. Suls and M. Chamberlain, "Psychological and physical stress induce differential effects on human colonic motility," The American Journal Of Gastroenterology, vol. 93, pp. 985-990, 1998.

[4] S. Cohen, R. Kessler and L. U. Gordon, "Strategies for measuring stress in studies of psychiatric and physical disorders," in Measuring Stress: A Guide For Health And Social Scientists, vol. 28, pp. 3-26, 1995.

[5] Y. Nakatake, H. Furuie, M. Yamada, H. Kuniishi, M. Ukezono, K. Yoshizawa and M. Yamada, "The effects of emotional stress are not identical to those of physical stress in mouse model of social defeat stress," Neuroscience Research, vol. 158, pp. 56-63, 2020.

[6] L. Kogler, V. Müller, A. Chang, S. Eickhoff, P. Fox, R. Gur and B. Derntl, "Psychosocial versus physiological stress—Meta-analyses on deactivations and activations of the neural correlates of stress reactions," Neuroimage, vol. 119, pp. 235-251, 2015.

[7] K. Hong, "Classification of emotional stress and physical stress using a multispectral based deep feature extraction model," Scientific Reports, vol. 13, p. 2693, 2023, doi: 10.1038/s41598-023-29903-3.

[8] J. Soederholm and M. Perdue, "II. Stress and intestinal barrier function," American Journal Of Physiology-Gastrointestinal And Liver Physiology, vol. 280, G7-G13, 2001, doi: 10.1152/ajpgi.2001.280.1.G7.

[9] A. Kavushansk, D. Ben-Shachar, G. Richter-Levin and E. Klein, "Physical stress differs from psychosocial stress in the pattern and time-course of behavioral responses, serum corticosterone and expression of plasticity-related genes in the rat," Stress, vol. 12, pp. 412-425, 2009.

[10] R. Costin, C. Rotariu and A. Pasarica, "Mental stress detection using heart rate variability and morphologic variability of EeG signals," The 2012 International Conference And Exposition On Electrical And Power Engineering, pp. 591-596, 2012, doi: 10.1109/ICEPE.2012.6463870.

[11] J. Rousselle, J. Blascovich and R. Kelsey, "Cardiorespiratory response under combined psychological and exercise stress," International Journal Of Psychophysiology, vol. 20, pp. 49-58, 1995.

[12] Y. Can, N. Chalabianloo, D. Ekiz and C. Ersoy, "Continuous Stress Detection Using Wearable Sensors in Real Life: Algorithmic Programming Contest Case Study," Sensors, vol. 19, p. 1849, Aug. 2019.

[13] R. Costin, C. Rotariu and A. Pasarica, "Mental stress detection using heart rate variability and morphologic variability of EeG signals," 2012 International Conference And Exposition On Electrical And Power Engineering, pp. 591-596, 2012.

[14] R. Garg, J. Santhosh, A. Dengel and S. Ishimaru, "Stress detection by machine learning and wearable sensors," The 26th International Conference On Intelligent User Interfaces-Companion, pp. 43-45, April 2021, doi: 10.1145/3397482.3450732.

[15] K. Dalmeida and G. Masala, "HRV features as viable physiological markers for stress detection using wearable devices," in Sensors, vol. 21, p. 2873, 2021, doi: 10.3390/s21082873.

[16] R. Castaldo, L. Montesinos, P. Melillo, C. James and L. Pecchia, "Ultra-short term HRV features as surrogates of short term HRV: A case study on mental stress detection in real life," BMC Medical Informatics And Decision Making, vol. 19, pp. 1-13, 2019.

[17] P. Gomes, P. Margaritoff and H. Silva, "pyHRV: Development and evaluation of an open-source python toolbox for heart rate variability (HRV)," Proc. Int'l Conf. On Electrical, Electronic And Computing Engineering (IcETRAN)," pp. 822-828, 2019.

[18] T. Iqbal, A. Simpkin, D. Roshan, N. Glynn, J. Killilea, J. Walsh, G. Molloy, S. Ganly, H. Ryman, E. Coen, A. Elahi, W. Wijns and A. Shahzad, "Stress Monitoring Using Wearable Sensors: A Pilot Study and Stress-Predict Dataset," Sensors, vol. 22, p. 8135, 2022.

[19] W. Rosenberg, M. Hoting and D. Mandic, "A physiology based model of heart rate variability," Biomedical Engineering Letters, vol. 9, pp. 425-434, 2019, doi: 10.1007/s13534-019-00124-w.

[20] F. Pedregosa, G. Varoquaux, A. Gramfort, V. Michel, B. Thirion, O. Grisel, M. Blondel, P. Prettenhofer, R. Weiss, V. Dubourg, J. Vanderplas, A. Passos, D. Cournapeau, M. Brucher, M. Perrot and E. Duchesnay, "Scikit-learn: Machine Learning in Python," Journal Of Machine Learning Research, vol. 12, pp. 2825-2830, 2011.

[21] J. Montiel, M. Halford, S. Mastelini, G. Bolmier, R. Sourty, R. Vaysse, A. Zouitine, H. Gomes, J. Read, T. Abdessalem, A. Bifet, "River: machine learning for streaming data in Python," The Journal of Machine Learning Research, vol. 22.1, pp. 4945-4952, 2021.

[22] B. Lakshminarayanan, D. Roy and Y. Teh, "Mondrian forests: Efficient online random forests," Advances In Neural Information Processing Systems, vol. 27, 2014, pp. 3140-3148, doi: 10.48550/arXiv.1406.2673.
[23] J. Healey and R. Picard, "Detecting stress during real-world driving tasks using physiological sensors," IEEE Transactions On Intelligent Transportation Systems, vol. 6, pp. 156-166, 2005.
[24] S. Ishaque, N. Khan, and S. Krishnan, "Trends in heart-rate variability signal analysis," Frontiers In Digital Health, vol. 3, p. 639444, Feb. 2021.

Developing Context-Based Applications Using Visual Programming

Case Studies on Mobile Apps and Humanoid Robot Applications

Martin Zimmermann
Department of Economics
Offenburg University
Offenburg, Germany
e-mail: m.zimmermann@hs-offenburg.de

Abstract— **Sensors and actuators enable creation of context-aware applications in which applications can discover and take advantage of contextual information, such as user location, nearby people and objects. In this work, we use a general context definition, which can be applied to various devices, e.g., robots and mobile devices. Developing context-based software applications is considered as one of the most challenging application domains due to the sensors and actuators as part of a device. We introduce a new development approach for context-based applications by using use-case descriptions and Visual Programming Languages (VPL). The introduction of web-based VPLs, such as Scratch and Snap, has reinvigorated the usefulness of VPLs. We provide an in-depth discussion of our new VPL based method, a step by step development process to enable development of context-based applications. Two case studies illustrate how to apply our approach to different problem domains: Context-based mobile apps and context-based humanoid robot applications.**

Keywords-Context-based Services; Sensors; Actuators; Mobile Applications; Location-based Services; Robot Applications; Humanoid Robots; Visual Programming.

I. INTRODUCTION

The main privilege of context-aware applications is to provide tailored services by analyzing the environmental context, such as location, time, weather condition, and seasons, and adapting their functionality according to the changing situations in context data without explicit user interaction. For example, mobile devices can obtain the context information in various ways in order to provide more adaptable, flexible and user-friendly services. In case of a tourist app, a tourist would like to see relevant tourist attractions on a map together with distance information, depending on its current location. Human robots, require sensors to gather information about the conditions of the environment to allow the robot to make necessary decisions about its position or certain actions that the situation requires. As a consequence, context-aware applications can sense clues about the situational environment making applications more intelligent, adaptive, and personalized.

Sensors and actuators as part of devises enable creation of context-aware applications in which applications can discover and take advantage of contextual information, such as user location, nearby people and objects. A general definition of context was given by Dey and Abowd [1]: "Any information that can be used to characterize the situation of an entity. An entity is a person, place, or object that is considered relevant to the interaction between a user and an application, including the user and applications themselves."

Categories of context information that are practically significant are [1]:

- Environmental Context: Include all the surrounding environmental conditions of current location (like air quality, temperature, humidity, noise level and light condition).
- Temporal Context: Consists of temporal factor such as current time, date, and season of the year.
- Personal (identity) Context: Specifies user's characteristics and preferences like name, age, sex, contact number, user's hobbies and interest.
- Spatial context: Involves any information regarding to position of entity (person and object) for instance orientation, location, acceleration, speed.

Combination of spatial, temporal, activity and personal contexts makes the primary context to understand the current situation of entities, these types of contexts can response basic question about when, where, what, who.

Visual Programming Languages (VPL) and hybrid visual programming languages are considered to be innovative approaches to address the inherent complexity of developing programs [2][3]. In this work, we introduce an in-depth discussion of a new VPL based method, to enable even programming beginners the creation of context aware applications.

The rest of the paper is organized as follows: Section 2 introduces visual programming concepts, especially flow-based and object-oriented approaches. To illustrate how to apply our approach to different problem domains, context-based mobile apps and context based humanoid robot applications serve as case studies in Sections 3 and 4.

Finally, the limitations of the VPL approach, as well as directions for future research are presented in Section 5.

II. VISUAL PROGRAMMING

VPLs let users develop software programs by combining visual program elements, like sensor and actuator objects, loops or conditional statements rather than by specifying them textually [1].

A. *VPL Concepts*

A comprehensive analysis of various VPLs including the strengths and weaknesses of VPLs, as well as guidelines to choose the most suitable VPL for the task in hand is described in [2]. There are two popular categories of VPLs: Flow-based VPLs and object-oriented VPLs.

In case of object-oriented VPLs, visual program elements are based on an object-oriented paradigm, i.e., decomposition of a system into a number of entities called objects and then ties properties and function to these objects. An object's property can be accessed only by the functions associated with that object but functions of one object can access the function of other objects in the same cases using access specifiers. Figure 1 shows the visual elements for a simple function call, following the representation of visual elements used in MIT AppInventor [4]. Objects, method calls, arguments and results of method calls are represented by visual elements with different shapes and colors. Clicking a button, touching a map, and tilting the phone are examples for user-initiated events.

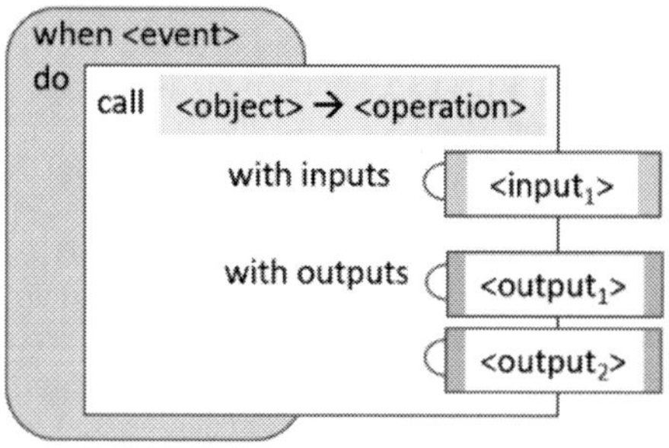

Figure 1. Object-oriented Visual Programming.

In general, flow-based VPLs offer different categories of elements, e.g., function calls, as well as control elements, like conditional statements. Programs are developed by placing them one after the other. For a parallel execution, more than one element may be used. Figure 2 illustrates the basic idea of a flow-based VPL-based program following the representation of visual elements used in Choregraphe [5]. First, the function f1 is called, then f2 and f3 are executed in parallel. Function elements are connected by using entry and exit ports. This is similar to BPMN, which stands for Business Process Model and Notation. BPMN is a standardized graphical modeling language used to represent and visualize business processes [6]. It is a widely adopted industry standard for modeling and documenting business processes, as it provides a consistent and easy-to-understand visual representation of a process. Flow elements are elements that connect with each other to form business workflows similarly to the visual building blocks of Choregraphe.

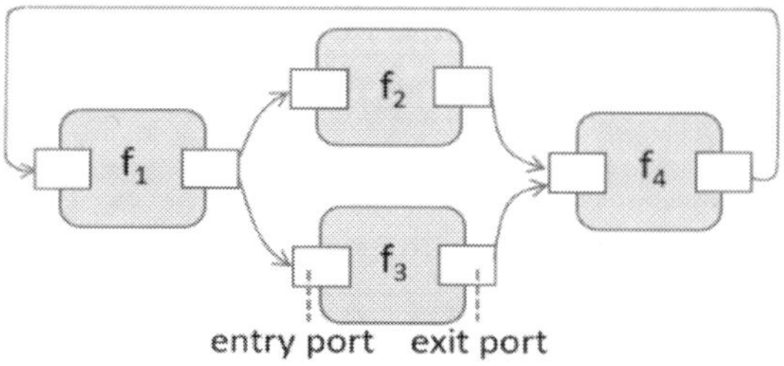

Figure 2. Flow-based Visual Programming.

Examples for flow-based systems are Flowgorithm [7], Microsoft VPL [8], and Choregraphe [5]. Flowgorithm is a general-purpose VPL that can be used to create flowchart representations of computing algorithms. It can translate a visual flowchart into eleven different textual programming languages, including C++ and Java. Flowgorithm is language independent but is not platform-independent; it is available only for Microsoft Windows operating system. Although it is not open source but is free for use.

Microsoft VPL is a programming environment based on graphical data-flows. It is aimed at engaging hobbyist programmers, as well as professionals. Furthermore, novice programmers can use it to learn programming, and experienced programmers can employ it for rapid prototyping. Microsoft VPL is a part of Microsoft Robotics Developers Studio (MRDS) used to develop software to guide robots. Choregraphe [5] represents also a flow-based visual programming environment which offers visual program elements for to easily develop complex robot applications, but Choregraphe also offers limited operations for the localization, mapping, as well as simple navigation functions.

B. *Use Case Driven Development*

Requirements engineering is done in two steps: Development of a use case diagram and specification of the use cases (each with input, output, steps). Based on the developed use cases, the VPL based application coding process is also use case driven and done in three steps:

- Development of a user interface for each use case (optional part).
- Selection of components (e.g., location sensor, QR code scanner, etc.).
- Flow-based or object-oriented visual programming

In Section 3, we show how our approach is applied to two very different problem domains, namely context-based mobile applications and robot applications.

III. Case Study: Humanoid Robot Applications

In this section, we present a case study of a context-based application for a humanoid robot. First, the sensors and actuators are analyzed, then the visual program elements are explained and finally the VPL based implementation of a context-based use case is described.

A. *Sensors and Actuators*

As an example, we use the popular humanoid robot Pepper (by SoftBank Robotics) [9][10], see Figure 3, a wheeled humanoid robot with sensors and actuators, i.e., torso, a head, two arms, with 20 degrees of freedom for motion in the body (17 joints for body language) and three omnidirectional navigation wheels to move around smoothly. It contains a set of various sensors to allow it to perceive objects and humans in its environment [11]:

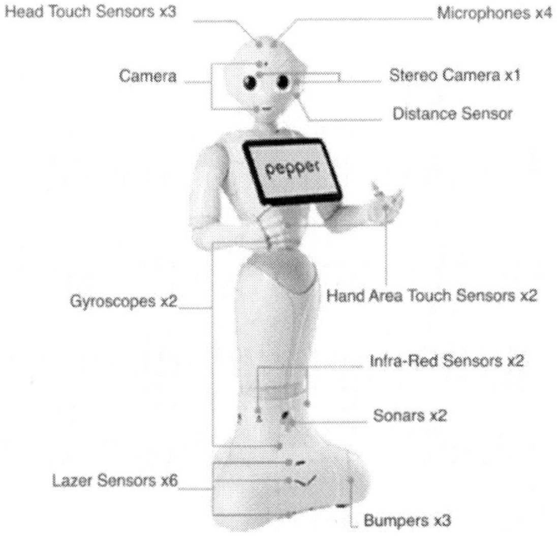

Figure 3. Sensors of the Pepper platform.

Pepper is able to localize a person talking to it, can distinguish multiple faces, determine eye contact or even recognize and react to basic emotions of the person it is talking with. Additionally, the humanoid robot is able to recognize someone's emotion not only by voice, but also by parameterizing facial expressions of interlocutors by using machine vision.

B. *Visual Program Elements*

Choregraphe is a visual programming environment for the Pepper platform. It offers visual program elements for different categories to easily develop complex robot applications, but Choregraphe also offers limited operations for the localization, mapping, as well as simple navigation functions [11].

Table I illustrates examples for the different categories of visual program elements, e.g., visual building blocks for speech creation, camera actions, or human face detection. Additional AI-based building blocks returns the gender, the age or the detected facial expression of the person in front of the robot. Building blocks include also logic functions and conditional statements, i.e., a condition and stimulate the then or else outputs depending on the boolean value of the condition.

TABLE I. Visual Program Element Examples.

Visual element	*Category*		
	Sensor	*Actuator*	*Other*
AnimatedSay (Animated Say)		Speech	
Face Detect (Face Detection)	Human Detection		
GetGender (Get Gender)	Human Understanding		
If statement (IF)			Logic

C. *Context-based Application: "Recognize and Greet People Scenario"*

In the following, we introduce a simple use case "Recognize & Greet People" as an example of a context-based application. The robot greets a person standing in front of it (after the robot has identified a face). Optionally (if questions are asked), the robot answers a question in the second use case "answer question" (not shown). The use case specification for the base use case is described in Table II. Based on a loop, four steps have to be executed: Detect face, determine gender, determine age and finally an animated say depending on the results of the previous blocks.

TABLE II. *Use Case: "Recognize & greet people".*

	Recognize and greet people
Input	Person in front of a robot
Steps	Loop 1: Detect face 2: Determine gender 3: Determine age 4: if (gender = female or gender = male) and (age < 30) Say "Hi, how can I help you" else if (gender = female) and (age >= 30) Say "Good Morning Madame, how can I help you" else if (gender = male) and (age >= 30) Say "Good Morning Sir, how can I help you"
Output	Voice output (depending on gender and age of face)

Based on the results of the requirements engineering, the implementation is also use case driven in three steps:

- Selection of visual program elements: Sensors, actuators and control elements
- Setting parameters of the visual programming building blocks (e.g., set the language of the "Animated Say" building block or the text to be spoken).
- Connection of visual programming building blocks, i.e., connecting their inputs and outputs, e.g., the output of "Face Detection" element must be connected to the input of the "Get Gender" element

Figure 4. Sensors and Actuators for the Use Case.

For the use case "Recognize & Greet People", the following sensors and actuators are required: Face Detection, GetAge, GetGender, and AnimatedSay (Figure 4). For example, GetGender returns the gender of the person in front of the robot. It is possible to set up the confidence threshold and the timeout.

Figure 5. Flow-based implementation of the Use Case.

Figure 5 illustrates the flow-based implementation of the use case.

IV. CASE STUDY: MOBILE APPLICATIONS

In this section, we present a case study of a location-based mobile application (spatial context). First, the sensors and actuators are analyzed, then the visual program elements are explained and finally the VPL based implementation of a concrete location-based service is described.

A. *Sensors and Actuators*

Sensors in mobile devices measure various environmental parameters, such as ambient air temperature and pressure, illumination, and humidity (Figure 6). This includes orientation sensors, magnetometers, but also barometers, photometers, and thermometers. Actuators mainly perform vibration-related functions, such as vibration and sound generation.

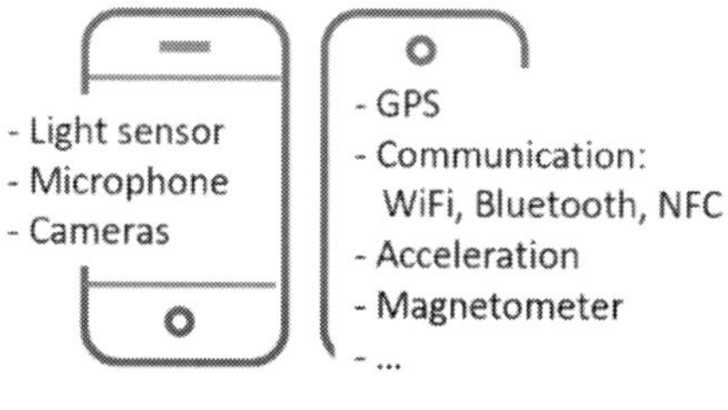

Figure 6. Sensors and Actuators of Mobile Devices.

B. *Visual Program Elements*

We use App Inventor [4] and Thunkable [12], which are both cloud-based visual development environments for mobile applications (Android and iOS). App Inventor and Thunkable provide the application developer with many different components to use while building a mobile app. Properties of these components such as color, font, speed, etc. can be changed by the developer. Available element categories are user interface elements, media, storage, location-based services etc. Elements can be clicked on and dragged onto the development screen area.

There are two main types of components: Visible and non-visible. Visible components such as buttons, text boxes, labels, etc. are part of the user interface whereas non-visible components such as the location sensor, QR Code scanner, sound, orientation sensor are not seen and thus not a part of the user interface screen, but they provide access to built-in functions of the mobile device.

TABLE III. VISUAL PROGRAM ELEMENT EXAMPLES.

Visual element	*Sensor*	*Actuator*	*Other*
Location Sensor	Get location information		
BLE	Connect to BLE devices		
Vibration		Vibrate for a specified time	
Loop			Iterate through elements

Table III illustrates examples for the different categories of visual building blocks. The location sensor provides location information, including longitude, latitude, altitude (if supported by the device), speed (if supported by the device). The Bluetooth Low Energy (BLE) component allows an application to find and connect to BLE devices and to communicate directly with them. The vibration actuator will vibrate the device for a specified time unit. The foreach element applies a set of functions to each element of a list, e.g., a list of tourist locations, part of a tourist app.

C. Context-based Application: Location-based Service Scenario

The use case in Table IV describes a location-based service [13] pattern in terms of input, output and steps to be executed. The template can be applied for instance to visualize some tourist attractions and the current position of a mobile user on a map. Filters are used to display certain tourist attractions, e.g., museums.

TABLE IV. USE CASE TEMPLATE "SHOW OBJECT(S)".

Use Case	***Show <object(s)> on a map***
Input	Filter
Steps	1: Determine the current geo position of the user 2: Show a map with the user's current position as the center point 3: Search the <object(s)> according to the specified filter (in a list of <object>) if found → Create marker(s) for the <object(s)>
Output	Map with markers: → marker for the current position of the user → marker(s) representing the <object(s)>

The implementation based on an object-oriented VPL, like MIT AppInventor, follows three steps:

- Selection of visual program building blocks: User-interface elements, sensors, actuators and control elements (e.g., if, switch elements).
- Definition of events (e.g., when button click, on map loaded)
- Calling methods (e.g., calling a method to get the current location of a user).

Event handler blocks specify how a program should respond to certain events. After, before, or when the event happens can all call different event handlers. There are two types of events: user-initiated and automatic.

Clicking a button, touching a map, and tilting the phone are user-initiated events. Sprites colliding with each other or with canvas edges are automatic events. Timer events are another type of automatic event. Sensor events function also as user-initiated events. For example, orientation sensor, accelerometer, and location sensor all have events that get called when the user moves the phone in a certain way or to a certain place.

Figure 7 shows the visual elements of the event-based programming part for the use case in Table IV, a simple location-based service. Objects, method calls, arguments and results of method calls are represented by visual elements with different shapes and colors. In a first step the current position of a user is determined by calling the method GetCurrentLocation. The resulting values (latitude and longitude) are used in the next step for the specification of the map center (two set operations). By calling the method addMarker, a marker is created in a third step. The arguments for the last method call are again visual elements (previously calculated values for the current latitude and longitude of the user).

Finally, a corresponding marker is generated for all objects of a list (by using a "for each item loop"), according to the specification in the use case description (Table IV).

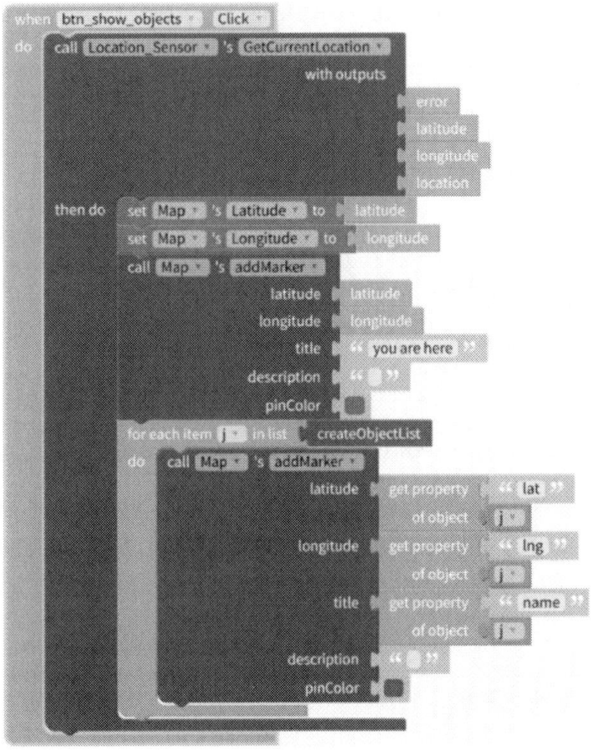

Figure 7. Implementation of a Use Case by Using an Object-Oriented VPL.

Creation of the object list could be based on a local list (as part of a mobile app) or a cloud-based object list. The creation / access to an object list in encapsulated in a separate function createObjectList. Finally, the objects behind a marker have to be visualized. For example, exhibits, like paintings in a gallery could be equipped with qr codes.

Object visualization in this case could mean to create a link to a video (painter explaining interesting background information) or a link to allow a tourist buying a print.

V. CONCLUSION

The main privilege of context-aware applications is to provide an effective, usable, rapid service by considering the environmental context (such as location, time, weather condition, and other attributes) and adapting their functionality according to the changing situations in context data. Use cases and visual programming are particularly well suited for programming beginners. However, visual programming environments are increasingly used in demanding problem domains, e.g., Internet Of Things (IoT) applications [14]. The development of use cases (in the sense of requirements engineering) as the starting point of an context-based application project has proven to be very advantageous.

Benefits of VPLs are short development times, low costs and increased efficiency and productivity [15]. VPL tools enable users with low technical skills to develop advanced software. Both VPL approaches, flow-based and object-oriented programming have their application areas. If parallel activities in particular are to be programmed, then flow-based systems are usually better suited. Our experience is that object-oriented VPLs seem to be more intuitive (especially for with low technical skills) due to the visual representation of objects, methods and the set and get operations.

A main drawback of the used programming environments is the identification and handing of runtime errors due to the lack of integrated debugging functions. However, our use case centered approach leads normally to manageable runtime error because each use case is developed and tested as a separate unit.

Future work will focus on the development of patterns and model-driven development [16]. Patterns are a well-known concept in the traditional software engineering. An architectural pattern is a general, reusable solution to a commonly occurring problem in software architecture. An architectural pattern becomes a reusable solution for a common set of problems in software development, addressing issues like high availability, performance, and risk minimization. Additionally, we focus on development of a repository of use case templates and visual code templates to improve design and implementation of context-based applications [17].

REFERENCES

[1] A. K. Dey and G. D. Abowd, "Towards a Better Understanding of Context and Contextawareness," CHI 2000 Workshop on The What, Who, Where, When, Why and How of Context-awareness, pp. 1–6, 2000.

[2] M. Idrees and F. Aslam, "A Comprehensive Survey and Analysis of Diverse Visual Programming Languages," VFAST Transactions on Software Engineering, vol.10, no. 2, pp. 47–60, 2022.

[3] R. Daskalov, G. Pashev, and S. Gaftandzhieva, "Hybrid Visual Programming Language Environment for Programming Training," TEM Journal, vol. 10 Issue 2, pp. 981–986, 2021.

[4] MIT App Inventor, https://appinventor.mit.edu, [retrieved: September, 2023].

[5] Choregraphe, http://doc.aldebaran.com/2-4/software/choregraphe/, [retrieved: October, 2023].

[6] BPMN, https://www.omg.org/bpmn, [retrieved: October, 2023].

[7] Flowgorithm, http://www.flowgorithm.org/, [retrieved: October, 2023].

[8] Microsoft vpl, https://msdn.microsoft.com/enus/library/bb483088.aspx, [retrieved: October, 2023].

[9] Pepper, http://doc.aldebaran.com/2-4/home_pepper.html [retrieved: October, 2023].

[10] C. Gómez, M. Mattamala, T. Resink, and J. Ruiz-Del-Solar, "Visual SLAM-Based Localization and Navigation for Service Robots": The Pepper Case. In Robot World Cup; Springer: Cham, Switzerland, pp. 32–44, 2018

[11] A. M. Marei, et al., "A SLAM-Based Localization and Navigation System for Social Robots: The Pepper Robot Case", Machines 2023, 11(2), pp. 47–60.

[12] Thunkable, https://thunkable.com, accessed: 2023-07-10.

[13] T D'Roza and G Bilchev, "An overview of location-based services," BT Technology Journal, vol. 21, no. 1, pp. 20–27, 2003

[14] M. Silva, J. P. Dias, A. Restivo, and H. S. Ferreira, "A Review on Visual Programming for Distributed Computation in IoT", Springer Nature Switzerland AG 2021, M. Paszynski et al. (Eds.): ICCS 2021, LNCS 12745, pp. 443–457, 2021

[15] D. Pinho, A. Aguiar, and V. Amaral, "What about the usability in low-code platforms? A systematic literature review", Journal of Computer Languages, Volume 74, pp. 1959–1981, 2023

[16] Md. Shamsujjoha, J. Grundy, Li Li, H. Khalajzadeh, and Q. Lu, Developing Mobile Applications Via Model Driven Development: A Systematic Literature Review, Information and Software Technology, Volume 140, December 2021.

[17] M. Zimmermann, "Location and Object-Based Mobile Applications", UBICOMM - International Conference on Mobile Ubiquitous Computing, Systems, Services and Technologies, pp. 34-39, 2023.

Agile and Reliable Design Decisions Based on the Perception of the Target Audience

Waumap Plataform: Methodological Tool to Involve Potential Consumers in the Design Process

Raquel Marzo Roselló
Instituto de Biomecánica
IBV
Valencia, Spain
e-mail: raquel.marzo@ibv.org

Adrián Colomer Granero
Universitat Politècnica de València
UPV
Valencia, Spain
e-mail: adcogra@upv.es

Abstract—**The conceptualization phase of a product plays a critical role in the design process as the decisions made during this phase directly affect the competitiveness of the product. During this phase, making decisions correctly becomes a very important activity for the company. However, decisions made during this phase often rely on subjective opinions, leading to uncertainty and errors, and consequently, high failure rates in the launch of new products. Few companies conduct user research studies due to the substantial time and cost investment, requisite expertise, technological resources, and access to adequately representative user samples necessary to obtain robust and reliable insights into people' preferences. This complexity makes such studies impractical for everyday design decision-making. In this context, this paper presents a methodological tool that allows companies to make objective decisions in the early phases of their design process. This tool takes advantage of the potential of Artificial Intelligence (AI) to analyze in a standardized, agile and autonomous way the perception of the consumer expressed naturally by a representative sample of remote users, combining classical qualitative user research with natural language processing techniques. The methodological tool has been validated through use cases from companies in different sectors in which remote tests have been carried out with representative samples of users, showing its suitability to obtain in a robust, reliable, agile and economical way the design optimization keys from the point of view of market acceptance.**

Keywords- market research solution; user insights; emotion design; decision-making; NLP.

I. Introduction

The conceptualization phase of a product plays a critical role in the design process. The decisions made during this phase directly affect the degree of innovation, quality of design solutions, costs involved, and overall competitiveness of the product. During this stage, companies continuously accept and reject options, making it critical to make informed and appropriate decisions [1]. However, decision-making during this phase largely relies on internal groups and collaborative sessions, which are highly subjective and influenced by the design team's perceptions, tastes, knowledge, and previous experiences [2]. This often leads to uncertainty and errors during the design process, which is evident from the high failure rates of Fast-Moving Consumer Goods (FMCG) launches. Nielsen [3] reports that 76% of FMCG launches fail in their first year of life, primarily due to the failure to address consumer needs or frustrations.

To address this issue, some companies conduct user research studies during the strategic definition phase to identify design requirements that cater to the needs of their target audience. However, these studies employ traditional market research methodologies, such as surveys, focus groups, and interviews, which are time-consuming and expensive. Furthermore, these studies are conducted at the beginning of the process, limiting the flexibility to respond to market changes in later stages of development.

Moreover, marketing and user research managers in companies face constraints in terms of time and resources [4], resulting in a reduced frequency of these studies. Delaying involving potential consumers until the final validation stage increases the cost of development modifications significantly. According to Forrester [5], the cost of fixing a design problem after launch is 100 times higher than if it had been identified and rectified in the early stages of development.

When studies are conducted during the design phase, companies typically employ cost-effective and swift techniques, such as surveys or AB tests. While these methods allow companies to decide on a design alternative among several options, they fail to provide information on the reasons for the preference or the related design elements, limiting the scope for maximizing customer satisfaction. In this context, there are user research techniques that focus not only on investigating user preferences but also on transferring them to the design elements of the product, such as Kansei Engineering, Conjoint analysis or the Repertory Grid Technique [6]. Another type of techniques used to design products are neuroscientific techniques, which allow us to know the influence of different attributes of the product (color, packaging...) on target audience' impact [7], being EEG and eyetracking the most used by researchers [8]. The drawback of this type of techniques is that they require slow and expensive research work with users, as well as an analysis and interpretation of complex results for the daily decision making of companies during their design process, an aspect that goes against the need to reduce launch times due to the continuous reduction of product life cycles.

For this reason, the development of a user-centered design methodological tool is planned to allow companies to make objective and reliable decisions in early phases of the product development process, replacing the most frequent approaches of decision-making based on intuition.

The main hypothesis of this research work was that the remote capture and subsequent automated analysis, employing AI algorithms, of the naturally expressed perception of a representative sample of potential users when faced with certain design alternatives, would obtain, in a simple and agile way for companies, key indicators for the optimization of the design in relation to market acceptance. While remote testing has inherent limitations, particularly the inability to present the physical product, it is deemed a favorable approach for approximating the capture of users' initial impressions during the early stages of development.

In order to achieve the proposed hypothesis, a methodological tool that incorporates AI has been developed using the "Customer Development" methodology [9], as explained in section II. Section III describes the results of 11 use cases from companies in different sectors and section IV, the reliability and robustness of the tool. Finally, the conclusions and future work of the study are shown in section V.

II. A New Methodological Tool for Including Potential Consumers in the Design Process

This section describes the tool developed, as well as the methodology followed for its development, considering the market problem.

A. *Final Design of the Methodological Tool for Design Decision Making*

To address the need for more efficient and effective methods for including users in the product development process, the Instituto de Biomecánica (IBV) has developed the Waumap methodological tool [13]. IBV has more than 40 years of experience providing advisory services to companies for designing target audience-oriented products using the People Oriented Innovation (IOP) methodology [6].

This tool helps companies to make design decisions in early phases. It allows companies to carry out design testing with a representative sample of their target audience, obtaining key indicators for optimizing the design from the point of view of maximizing its positive perception.

Waumap tool utilizes AI to analyze consumer perceptions of various design alternatives in a standardized, agile, and autonomous manner, allowing companies to make reliable and objective decisions during their design processes.

The Waumap study is conducted in three simple steps:

1. Defining the test based on two images or videos and the characteristics of the target audience. The test comprises a survey (with concepts defined by the company) and analysis of natural language using AI from open opinions expressed freely. Eye-tracking may also be incorporated.
2. Launching the remote study to a database of potential users that fit the defined profile.
3. Automatically generating a report, which can be received within 7 days of launching the study. Figure 1 shows an example of a Waumap report. The Waumap report provides information on the factors driving preference and the emotions evoked by each design alternative through opinion polarity analysis. The eye-tracking feature generates heat maps displaying the areas of greatest visual attention.

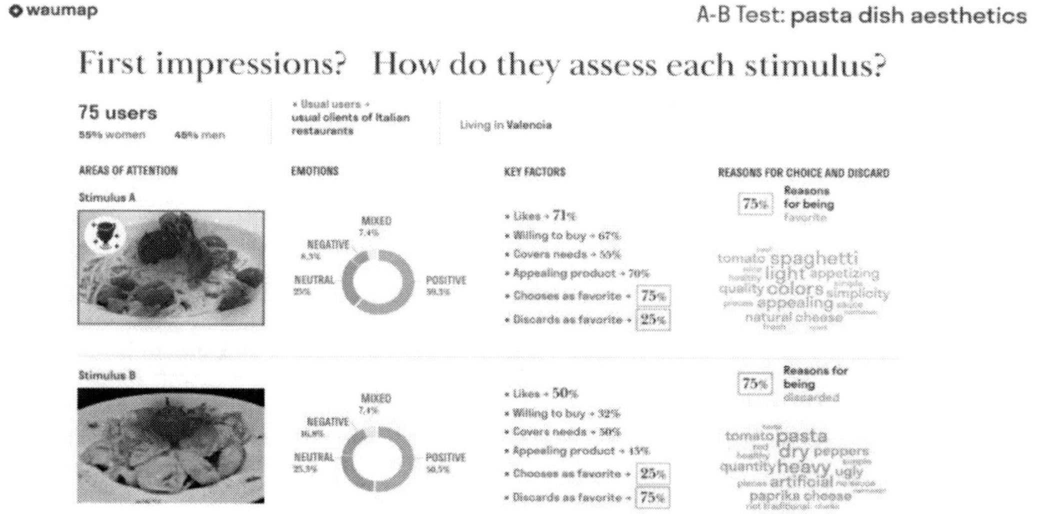

Figure 1. Waumap test AB eye tracking report

The table below provides a detailed description of the methodology to obtain the key variables of the Waumap report by automatically recording and analyzing user perceptions.

TABLE I. WAUMAP METHODOLOGY

Variables	Recording techniques	Analysis techniques
Areas of greatest visual attention	Gaze tracking recording via webcam.	IBV's own programming using OpenCV Models.
Emotions (positive-negative-mixed-neutral)	Recording of first impression expressed in a natural way, either by voice or text.	- Amazon Transcribe. Amazon Web Services (AWS) automatic speech recognition service that facilitates speech-to-text conversion. It provides the transcribed text, as well as confidence scores between 0 and 1 (self-assessments by the service on how well it may have transcribed the text). It is used to convert the first impression expressed naturally by voice into text. - Amazon Comprehend. AWS natural language processing (NLP) service that allows, using machine learning techniques, to analyze and understand the meaning of text in different languages. Specifically, the Waumap tool uses the "Detect Sentiment" function, which analyzes the text and determines the polarity of the sentiment, i.e. whether the text is positive, negative, neutral or mixed. This function returns an inference of the predominant sentiment and provides for each sentiment the probability between 0 to 1 that it has been correctly detected.
Key factors	Closed survey with concepts, to know the perception of the design alternative in some concepts predefined by the company. The ordinal Likert scale of 5 categories has been defined, as it is the most used scale in market research and easy to answer (Likert, 1932), being 1 the value associated with the most disagreement (e.g., I do not like it at all) and 5 the value associated with the most agreement (e.g., I like it a lot).	Calculation of the total % of participants with answers classified as positive grouped in Top 2 Box value, i.e., percentage of people who select for each survey concept the two most positive values.
Favorite design alternative	Prioritization of design alternatives, clicking on the favorite stimulus.	Calculation of % of participants selecting and discarding each design alternative as favorite.
Reasons for choice and discard	Justification of the reasons for the prioritization of design alternatives, expressed in a natural way by text.	- Natural Language Toolkit (NLTK). A set of Python libraries and programs that facilitate natural language processing, such as the "nltk.corpus" library which provides access to a variety of corpora (text datasets). Among them is "Stopwords" which contains a predefined list of common words that do not provide value for text compression, such as articles, prepositions and conjunctions. - Wordcloud. Python function that generates a visual representation of the words in a text, where the size of each word represents its frequency. Prior to such visualization, natural language processing is required, such as the elimination of common words with the "Stopwords" function. In the Waumap tool, word clouds are represented for the selection and discard reasons for each design.

B. *Definition of a Methodological Tool taking into account Market Needs*

The development of this tool was carried out using the "Customer Development" methodology, which aims to understand the market problem and validate that the proposed solution will indeed meet customer needs and demand, reducing business risks by testing hypothesis [9]. Firstly, hypothesis about its company profile and value proposition were defined, using the "Business Model Canvas" template, a visual chart with elements describing a new business model [10]. On the other hand, hypotheses about the companies' problem and the possible solution were specified. These hypotheses were classified following the model proposed by Josh Seiden [11], based on risk and uncertainty, for prioritization when validating with companies:

- Risk. If the hypothesis were false, the methodological tool would have a high risk of not achieving market acceptance. The classification based on risk has been carried out at expert criterion.
- Uncertainty. A hypothesis with high uncertainty is a hypothesis with a high unknown (lack of data) regarding its truth or falsity. To classify the hypotheses based on uncertainty, a state of the art on decision making in the design process of companies has been carried out.

After the definition and prioritization of the hypotheses related to market acceptance, test cards [11] were developed for the definition of the experiments to be carried out to validate the most critical hypotheses and with greater uncertainty. In the experiments, Minimum Viable Products (MVP) of Waumap tool were used.

To validate these hypotheses, from the beginning of the research work, a sample of 11 Valencian companies with variability in sectors and type of products (food, distribution, home, clothing, cleaning, advertising and tourism) collaborated and were interviewed after the experiments. The interviews lasted 30 minutes and followed the guidelines recommended in the book "Lean Customer Development" [9].

Table II shows the results of the hypothesis validation.

TABLE II. RESULTS OF THE HYPOTHESIS VALIDATION

Hypothesis	Experiment results
The Waumap results are useful for its decision making	True. 8 of the 11 companies (> 60%) affirmed that waumap results are credible and clear for its decision making.
Waumap tests help the company in its decision-making processes	True. 10 of the 11 companies (> 60%) named design problems of their company to test with waumap.
Companies have adequate material to evaluate different design alternatives in Waumap	True. 8 of the 11 companies (>60%) provided stimuli of a design problem relevant to their company
Waumap fits into the company's daily taks (actions to be carried out, availability...)	True 8 of the 11 companies (> 60%) completed the configuration process of a Waumap test and valued its ease of use.

Another parameter to be defined prior to the final development of the tool was the sample size. To optimize the number of users needed to obtain reliable conclusions, a pilot study with 5 Waumap tests was conducted to estimate the effect size of the variable "favorite design alternative" (see Table I). Table III shows the value of this variable in the 5 tests.

TABLE III. DIFFERENCES BETWEEN THE PREFERENCE OF DESIGN ALTERNATIVES

Waumap Tests	Favorite design alternative	Preference differences
Casual footwear Test	Stimulus A by 52%	4% (52%-48%)
Advertising campaigns Test	Stimulus A by 57%	14% (57%-43%)
Cleaning products Test	**Stimulus A by 65%**	**30%** (65%-35%)
Hotel rooms Test	Stimulus A by 75%	50% (75%-25%)
Pasta dishes Test	Stimulus A by 75%	50% (75%-25%)

After obtaining the results of the tests, it was deemed appropriate to establish an objective with the aim of detecting substantial differentials exceeding 30% in preference among design alternatives. This objective pertains, in particular, to the assessment of designs conducted in the context of the "cleaning products Test", which serves as a representative archetype of the design category slated for evaluation using the Waumap tool, primarily intended for novel product concept assessment.

The appropriate formula [12] with the data of "cleaning products test" was applied for experimental designs (contrast test) in which non-parametric statistical tests are applied (X^2).

$$W = \sqrt{\sum_{i=1}^{m} \frac{(P_{1i} - P_{0i})^2}{P_{0i}}} = \sqrt{\frac{(0.65 - 0.5)^2}{0.5} + \frac{(0.35 - 0.5)^2}{0.5}} = 0.3$$

Cohen's tables [4] were consulted for a test power of 75%, α=0.05 (95% confidence level), u=1 (2 designs alternatives -1) and W=0,3, obtaining n=75 users.

Power of χ^2 test at a = .05, u = 1

	w								
N	.10	.20	.30	.40	.50	.60	.70	.80	.90
25	08	17	32	52	70	85	94	98	99
30	08	19	38	59	78	91	97	99	*
35	09	22	43	66	84	94	99	*	
40	10	24	47	71	89	97	99		
45	10	27	52	76	92	98	*		
50	11	29	56	81	94	99			
60	12	34	64	87	97	*			
70	13	39	71	92	99				
80	15	43	76	95	99				
90	16	47	81	97	*				
100	17	52	85	98					
120	19	59	91	99					
140	22	66	94	*					
160	24	71	97						
180	27	76	98						
200	29	81	99						
250	35	89	*						
300	41	93							
350	46	96							
400	52	98							
500	61	99							
600	69	*							
700	75								
800	81								
900	85								
1000	89								

Figure 2. Cohen tables for sample size stimation.

III. USER COMPANIES' FEEDBACK OF THE WAUMAP TOOL FOR PRODUCT DEVELOPMENT

After conducting the "Customer Development" methodology with the collaborating companies, the potential company profile of the methodological tool, and their needs and demands were identified (Table IV).

TABLE IV. COMPANY PROFILE, NEEDS AND DEMANDS

Sectors	Food, distribution and cleaning. Tourism and advertising.
Main activity	Development of consumer goods, excluding commodity products. Tourism and advertising services.
Other features	• High billing • High investment in R&D before product launch or high product rotation. • Restless, eager to innovate

	• Currently carrying out some testing before launch, even if it is internally. • Awared that company could improve its decision making in the design.
Needs/problems	• Most companies claim to currently spend a lot of time and effort making decisions in the design and launch of products. • They have a lot of uncertainty about the success of new product launches (a company claims that 9 out of 10 products fail). • In general, companies say they do not have the budget to carry out market studies that allow them to predict market acceptance before launch (in addition, these types of studies usually have long deadlines and do not fit into their processes).
Demands/solutions	Companies would like a solution that would allow them... • Reduce the risk of development due to non-acceptance by the market. • Decide/discriminate between different design alternatives. • Know the value proposition of their products/services and also the points for improvement.

Furthermore, the Waumap methodological tool has been validated through various use cases with the 11 collaborating companies. These cases involved remote testing with representative user samples, which demonstrated the tool's effectiveness in obtaining the keys to design optimization. Companies appreciate the tool's simplicity, intuitiveness, and agility, especially in the areas of new product concepts, supermarket shelf (physical and/or virtual), packaging, and corporate image.

The Waumap results have provided valuable insights in the decision-making process, allowing companies to access a representative sample of users remotely and reduce costs in user recruitment, testing, and prototype development. Companies also appreciate the short period of time it takes to obtain results, with reports typically available within a week. The report's clarity and ease of understanding not only helps companies make decisions but also justifies them internally and to B2B clients.

The preference keys obtained through natural language processing analysis provide the most value in decision-making, enabling companies to understand the reasons for preference/rejection and their relationship with the design elements.

IV. Realiability and Robustness of Waumap

This section describes the conducted analyses to show the suitability of the tool for obtaining design optimization keys in a reliable and robust manner.

A. *Reliability of Waumap Tool*

To demonstrate the reliability of the tool, a comparative analysis of the results obtained through a test developed with the Waumap tool and through a classic market research study (two focus groups) was carried out. In these studies, two sauce labels were assessed as design alternatives (Figure 3).

It is worth mentioning that the focus groups were carried out by one of the collaborating companies (Chovi), as this was the methodology they commonly used for this type of design problems and the results were not shared until Waumap test was completed.

Figure 3. Design alternatives evaluated in the comparative study.

The results obtained with both tests were identical, obtaining the same design as a favorite and the same reasons for choice and discard.

The comparative study has made it possible not only to demonstrate the reliability of the results, but also to demonstrate the advantages of Waumap tool over traditional studies, as the results are obtained in a much more agile, simple and economical way (see Table V).

TABLE V. Comparison between Waumap Test and Focus Groups

Waumap Test	2 focus groups
N=75 users	N=14 users
Reporting deadline: 3 days	Reporting deadline: 3 weeks
Staff hours: 2,5 hours	Staff hours: 45 hours
User gratification cost: 225 euros	User gratification cost: 420 euros

B. *Robustness of Waumap Tool*

To demonstrate the robustness of the tool, the perception results of a sauce packaging, assessed in two different Waumap tests with different samples of users and in different time periods were compared (Figure 4).

Figure 4. Design evaluated in two Waumap tests.

The results related to the first impressions expressed naturally extracted by natural language processing algorithms were similar: polarity of the sentiments and word clouds.

V. Conclusion and Future Work

The analysis of the perception of the consumers who have participated in the Waumap' use cases have allowed to show that the automatic analysis of the user's perception through AI reduces the time and effort of analysis to generate the keys that make it easier for companies to make more informed decisions in the design of their products.

The Waumap tool has a positive impact on the decision-making process, facilitating better integration of the target audience in the product conceptualization phase by a simple, intuitive and agile process. Furthermore, this methodological tool provides key insights into design optimization from the perspective of market acceptance, with particular utility evident in the following application domains: new product concepts, supermarket shelf (physical and/or virtual), packaging and corporate image.

The following future lines of work have been identified:

- Including more than two design alternatives in the Waumap test.
- Exploring sentiment analysis by also analyzing tone of voice.
- Obtaining and validating a preference prediction model based on sentiment polarity, when more data is obtained with future use cases.

Acknowledgment

The authors would like to acknowledge the Directorate General of Innovation of the Ministry of Innovation, Universities, Science and Digital Society, who financed the DEMOCRATIOP project (CONV21/DGINN/10 and CONV22/DGINN/07). Also, we thank the following companies: Choví, Costa Brava, Pikolinos, Hosbec, SPB, Playfilm, Peronda and Luanvi. The development and validation of the Waumap tool would not have been possible without the active participation of these companies. Finally, we would like to thank José Antonio de Miguel Visa for his advice on the development of the Waumap tool using agile methodology and Lean Customer Development.

References

[1] J. C. Bernal, "Study on the decision-making process during the conceptual phase of product design". Iconofacto, 2016.

[2] J. Rasmussen, A. Pejtersen and L. P. Goodstein, "Cognitive engineering systems". New York: John Wiley & Sons, 1994.

[3] R. Nielsen, "Nielsen Breakthrough Innovation Report", European Edition, 2014.

[4] I. Bortels, "Are we facing a 'revival' of qualitative research?" I&A. Investigación y Marketing, 2019.

[5] R. K. Souza, "Get ROI from Design". Cambridge: Forrester Research, 2001.

[6] Instituto de Biomecánica, New techniques for the development of innovative user-oriented products, 2001.

[7] K. Ploom, K. Pentus, A. Kuusik and U. Varblane, "The effect of culture on the perception of product packaging: a multimethod cross-cultural study". J. Int. Cons. Market. 32, 1–15, 2019.

[8] L. Alvino, "Picking Your Brains: Where and How Neuroscience Tools Can Enhance Marketing Research". Frontiers in Neuroscience, 2020.

[9] C. Alvarez, Lean Customer Development: Building Products Your Customers Will Buy, 2017.

[10] A. Osterwalder, Value Proposition Design - How to Create Products and Services Customers Want, 2014.

[11] J. Seiden, Lean UX: Designing Great Products with Agile Teams, 2016.

[12] N. Cohen, Statistical Power Analysis for the behavioral Sciences. Academic Press, Inc., 1977.

[13] https://waumap.ibv.org/, retrieved on 2023.